MAKING
Sense
of Statistics

Making Sense of Statistics, Eighth Edition, is the ideal introduction to the concepts of descriptive and inferential statistics for students undertaking their first research project. It presents each statistical concept in a series of short steps, then uses worked examples and exercises to enable students to apply their own learning.

It focuses on presenting the "why," as well as the "how" of statistical concepts, rather than computations and formulas. As such, it is suitable for students from all disciplines regardless of mathematical background. Only statistical techniques that are almost universally included in introductory statistics courses, and widely reported in journals, have been included.

This conceptual book is useful for all study levels, from undergraduate to doctoral level across disciplines. Once students understand and feel comfortable with the statistics presented in this book, they should find it easy to master additional statistical concepts.

New to the Eighth Edition

- ❏ Reorganization of chapters to allow better progress in conceptual understanding
- ❏ Additional discussions on program evaluation, displays of outcomes, and examples
- ❏ Chapter objectives at the beginning of each chapter are listed with clear learning objectives for the reader

- ❏ Expanded appendices include a reference to common computational formulas and examples
- ❏ A glossary of key terms has been updated to function as a useful vocabulary list for use in a first course in statistics
- ❏ Updated online resources, including a basic math review and answers, PowerPoint slides, and a test bank of questions

The downloadable Support Material can be accessed at: www.routledge.com/9781032289649

Deborah Mikyo Oh has been Professor in Research Methods and Statistics at California State University, Los Angeles, USA, since 1998. She earned her PhD from Columbia University, New York, USA, and teaches statistics to a diverse student population at undergraduate, graduate, and doctoral levels.

MAKING Sense of Statistics

A Conceptual Overview

Eighth Edition

Deborah Mikyo Oh

and Fred Pyrczak

Routledge
Taylor & Francis Group

NEW YORK AND LONDON

Eighth edition published 2023

by Routledge
605 Third Avenue, New York, NY 10158

and by Routledge
4 Park Square, Milton Park, Abingdon, Oxon, OX14 4RN

Routledge is an imprint of the Taylor & Francis Group, an informa business

First edition published by Pyrczak Publishing 1995

Seventh edition published by Routledge 2018

Library of Congress Cataloging-in-Publication Data
A catalog record has been requested for this book

ISBN: 978-1-032-28962-5 (hbk)
ISBN: 978-1-032-28964-9 (pbk)
ISBN: 978-1-003-29935-6 (ebk)

DOI: 10.4324/9781003299356

Typeset in Warnock Pro
by Deanta Global Publishing Services, Chennai, India

Access the Support Material: www.routledge.com/9781032289649

Contents

Introduction to the Eighth Edition   vii

Introduction: What Is Research?   xi

PART A
The Research Context   1

Chapter 1	The Empirical Approach to Knowledge	3
Chapter 2	Types of Empirical Research	9
Chapter 3	Scales of Measurement	15
Chapter 4	Descriptive, Correlational, and Inferential Statistics	21

PART B
Sampling   25

Chapter 5	Introduction to Sampling	27
Chapter 6	Random Sampling	33
Chapter 7	Sample Size	41
Chapter 8	Standard Error of the Mean and Central Limit Theorem	49

PART C
Descriptive Statistics   57

Chapter 9	Frequencies, Percentages, and Proportions	59
Chapter 10	Shapes of Distributions	63
Chapter 11	The Mean: An Average	69
Chapter 12	Mean, Median, and Mode	73
Chapter 13	Range and Interquartile Range	79
Chapter 14	Standard Deviation	85
Chapter 15	*z* Score	91

Contents

PART D

Correlational Statistics   97

Chapter 16	Correlation	99
Chapter 17	Pearson r	105
Chapter 18	Scattergram	111
Chapter 19	Coefficient of Determination	121
Chapter 20	Multiple Correlation	127

PART E

Inferential Statistics   133

Chapter 21	Introduction to Hypothesis Testing	135
Chapter 22	Decisions about the Null Hypothesis	141
Chapter 23	Limitations and Implications of Significance Testing	147

PART F

Means Comparison   153

Chapter 24	Introduction to the t Test	155
Chapter 25	Independent Samples t Test	163
Chapter 26	Dependent Samples t Test	169
Chapter 27	One-Sample t Test	175
Chapter 28	Reporting the Results of t Tests: Display of Outcomes	179
Chapter 29	One-Way ANOVA	185
Chapter 30	Two-Way ANOVA	193

PART G

Predictive Significance   203

Chapter 31	Chi-Square Test	205
Chapter 32	Effect Size	211
Chapter 33	Simple and Multiple Linear Regression	217

Appendix A: Computations	223
Appendix B: Notes on Interpreting Pearson r and Linear Regression	231
Appendix C: Table of Random Numbers	233
Comprehensive Review Questions	235
Glossary	257
Index	267

Introduction to the Eighth Edition

Four types of students need a conceptual overview of statistics:

1. Students preparing to be consumers of empirical research who need to understand basic concepts in order to interpret the statistics reported by others in journals, at conferences, and in reports prepared by their supervisors and co-workers.
2. Students taking a traditional introductory statistics course who are getting lost in the details of deriving formulas, computing statistics, and sorting through an overload of theory. For them, this book is an ideal supplement.
3. Students taking an introductory statistics course in which computers are used for all major calculations and who need a text that helps them understand the meaning of their output. For them, this book may be used as their main textbook.
4. Students who have taken a statistics course but need to review essential concepts in preparation for a subsequent course for which statistics is a prerequisite, or in preparation for their theses or dissertations.

Coverage

The coverage is selective. To be included in this book, a statistical technique had to be one that is almost universally included in introductory statistics courses and widely reported in journals. Once students understand and feel comfortable with the statistics that meet these criteria, they should find it easier to master additional statistical concepts.

Computations and Formulas

You will find very few formulas and computational procedures described in the body of this book, only to the extent that these add to your conceptual understanding. We have assumed that for some students, a non-computational approach is best for developing an understanding of the meaning of statistics. For students who are becoming consumers of research, this may be all that is needed. For others, computers can handle the computations; such students need to know which statistics are available for various purposes and how to interpret them. Some computational formulas have been added to Appendix A in this edition as a resource for further computational understanding.

Steps in Using This Book

To get the most from this book, follow these steps: (1) read a chapter while ignoring all references to footnotes and appendices because these contain information that may distract you as you master essential concepts; (2) read the chapter again, pausing to read the footnotes and appendices when you encounter references to them; (3) read the summary statements in the sidebars for review; and (4) answer the exercise questions at the end of the chapter.

If you have not studied statistics before, you will find that most of the concepts in this book are entirely new to you. It is not realistic to expect to skim the text once to achieve full mastery. In addition, pilot tests indicated that students who read each chapter twice before answering the end-of-chapter questions mastered the concepts more quickly than those who read it only once. The latter did more "hunting and pecking" through the text to look for answers, a slow process that does not provide an overview and leads to errors. Thus, you should save time and achieve greater mastery by following the recommended steps faithfully.

Five multiple-choice items for each chapter of this book are included near the end, in the Comprehensive Review Questions. While these were designed to help students prepare for midterm and comprehensive final exams, they may be used at any time during the course when a review is needed.

New to the Eighth Edition

Changes in the Eighth Edition include reorganizing the chapters and putting correlational statistics in one section. Additional discussions on program evaluation, displays of outcomes, and examples have been added for better understanding. Chapter objectives at the beginning of each chapter are listed with clear learning objectives for the reader. Expanded appendices include a brief reference list of some common computational formulas and examples. At the end of the book, a glossary of key terms has been updated with references to chapters in parentheses.

Acknowledgments

We are grateful to Dr. Robert Morman and Dr. Patricia Bates Simun, California State University, Los Angeles; to Dr. Roger A. Stewart, University of Wyoming; to Dr. Richard Rasor, American River College; and to Dr. Matthew Giblin, Southern Illinois University, Carbondale, for their helpful comments on the various drafts of this book. Errors and omissions remain the responsibility of the authors.

Fred Pyrczak
Deborah M. Oh

Introduction: What Is Research?

Research involves coming to know something using a scientific method, which is a systematic way of acquiring new knowledge.

In life, we make observations based on our own experiences and draw decisions or conclusions. For example, we might observe that my dog, Fluffy, barks and jumps at strangers and conclude that all dogs bark and jump at strangers.

The basic way of knowing something involves observations through the sensory system: sight, smell, taste, hearing, and touch. But the sensory system is not a reliable basis from which to draw conclusions because it can be misleading and subjective. What we feel may not be accurate. The next order of coming to know something is through an agreement with others. If several people agree that a particular mountain is too steep to climb, we cannot conclude that the mountain is impossible to climb. It is hard enough to agree with others, but an agreement or the majority voice does not always speak the truth.

How about consulting an expert before a decision or conclusion is made? We can consult an expert, but even trusted sources come with personal biases. How about logic? Logic

Research involves using a scientific method of acquiring new knowledge.

has its own limitations. We know that not all conclusions that seem logical may be right. Our sensory system, an agreement with others, an expert's consultation and logic are subjective and unreliable methods of gaining knowledge. This brings us to the most reliable method of gaining new knowledge, an empirical approach, using a scientific method to study a research problem.

The scientific method involves identifying, formulating, determining, organizing, and interpreting.

The *scientific method* of gaining knowledge generally follows five parts: (1) identifying a problem, (2) formulating hypotheses, (3) determining what information needs to be collected, (4) organizing the gathered information, and then (5) interpreting the results. Then, based on this new knowledge, the cycle repeats again back to identifying a problem based on what we already know, thereby increasing our knowledge over time through repeated cycles.

The goal of research is to know by describing, predicting, and explaining situations involving human beings.

The goal of research then is to know by describing, predicting, and explaining situations involving human beings using an empirical approach, which is a formal and systematic way of applying these steps of a scientific method of research to study a problem.

Quantitative research uses numerical data to determine a cause-and-effect relationship.

Research is broadly categorized into quantitative and qualitative. The two categories are similar in that both involve collecting and analyzing data, as well as arriving at conclusions and interpretations. *Quantitative research* uses numerical data to determine a cause-and-effect relationship. *Qualitative research* uses verbal narrative data to describe how things are and what they mean.

Qualitative research uses verbal narrative data to describe how things are and what they mean.

Here are some examples of various quantitative and qualitative research questions.

EXAMPLE 1
About students with ADHD: *Quantitative research question*: What are the differences in the number of discipline referrals recorded between boys and girls with ADHD? *Qualitative research question*: What are the social experiences of students with ADHD in middle school?

EXAMPLE 2

About first-generation college students:

Quantitative research: What factors best predict first-generation college students to graduate successfully?

Qualitative research: What challenges do first-generation college students experience in their first year in college?

Verbal narrative data in a qualitative research would provide in-depth information and be descriptive in its conclusion, but data analysis is time-consuming. Numerical data in a quantitative research is easier to analyze and yields conclusive results. However, much of the in-depth information is lost in numbers. Despite these advantages and disadvantages, both categories of research are useful for various purposes.

What goes into a typical research study? A typical research study in an outline form includes the following: *Introduction*, where the topic is generally described; *Literature Review* to describe what we already know about the topic at hand; *Methodology to* describe the procedure of the study; *Results to* report on the statistical analyses of the gathered information; then finally, *Discussion/Conclusion* to interpret the results to answer the research question at hand.

> A typical research study includes: *Introduction, Literature Review, Methodology, Results, and Discussion/ Conclusion.*

One must understand, however, the limitations of using an empirical approach to gain knowledge, in that it cannot answer all questions. Additionally, not all things can be measurable, especially those of a philosophical, ethical, emotional, or spiritual nature; what is measurable may not capture the full depths of the context. Finally, the measurement tools may be limiting or be in error.

The purpose of this book is to introduce statistics as a *tool* to understand the world around us using quantitative (numerical) data. Thus, the book discusses the research context and sampling in Part A and Part B as the research context of statistics. Then the conceptual overview of statistics as a tool is discussed in the remaining parts, C through G, for understanding how to analyze quantitative data.

> The purpose of this book is to introduce statistics as a *tool* to understand the world around us using quantitative (numerical) data.

THE RESEARCH CONTEXT

The research context begins with the basic concept of the empirical approach and the terminologies that follow. Part A introduces this concept, as well as various types of research. Furthermore, the distinction between quantitative and qualitative approaches to research is also described in depth. In addition, scales of measurement are introduced. This part defines what *variables* are and discusses the four types of measurement. And finally, a broad-stroke overview of descriptive, correlational, and inferential statistics is discussed before they are described in depth in subsequent chapters. The following is the list of topics discussed in Part A: The Research Context.

Chapter 1. The Empirical Approach to Knowledge
Chapter 2. Types of Empirical Research
Chapter 3. Scales of Measurement
Chapter 4. Descriptive, Correlational, and Inferential Statistics

DOI: 10.4324/9781003299356-1

The Empirical Approach to Knowledge

Chapter Objectives

The reader will be able to:

- ❏ Explain that the empirical approach or scientific method refers to acquiring knowledge based on direct observations.
- ❏ Identify that empirical research requires advance planning.
- ❏ Recognize that empirical research begins with a research question(s).

Empiricism refers to using direct observation to obtain knowledge. Thus, the ***empirical approach*** or *scientific method* to acquire knowledge is based on making observations of individuals or objects of interest.[1] Note that making ***everyday observations*** is an application of the empirical approach. For instance, if we observe that a traffic officer regularly hides behind the bushes at a certain intersection and frequently issues tickets at that location, we might say we *know* that someone who runs a stop sign at that intersection is likely to get a ticket. Unfortunately, generalizations based on everyday observations are often misleading. Here is an example.

Empiricism involves direct observation.

The ***empirical approach*** is based on observation.

Everyday observation is an application of the empirical approach.

DOI: 10.4324/9781003299356-2

Generalizations based on everyday observations are often misleading.

Suppose you observe that most of your friends and acquaintances plan to vote in favor of a school bond measure to build new schools. Unless they are a good cross section of the electorate, which is unlikely, you may be wrong if you generalize to the population of voters and predict that the measure will pass in the election.

Empirical research is planned in advance.

Most researchers systematically use the empirical approach to acquire knowledge. When they do, we say that they are engaging in **empirical research**. A major distinction between empirical research and everyday observation is that empirical research is planned in advance.

Researchers plan *what, whom, how, when,* and *under what circumstances* to observe in order to answer the questions.

Based on a theory or hunch, researchers develop research questions. And then researchers plan *what, whom, how, when,* and *under what circumstances* to observe in order to answer the questions. When a researcher predicts the answer to a research question prior to conducting research, we say that he or she has a *hypothesis*. In other words, a **hypothesis** is a prediction of the outcome of research.

Hypothesis is a prediction of the outcome of research.

1. They plan *what* to observe. What does a researcher want to know? Research questions are the heart of all empirical research. It could be the effects of public stigma and other perceived barriers on mentally ill patients, or how teachers feel they are making a difference through teaching.

Planning how to draw a **sample** from a **population** is important in conducting valid research.

2. They plan *who* (or *what*) to observe. Who or what does a researcher want to study? It could be all mentally ill patients in a hospital ward or all public school teachers in Pennsylvania. These are known as **populations**. A population consists of groups of interest in one's research study. When a population is large, researchers often plan to observe only a **sample** (i.e., a subset of a population). Planning how to draw an adequate sample is, of course, important in conducting valid research.[2]

Measuring instruments are constructed or selected.

3. They plan *how* to observe. How will a researcher get the information needed to answer the research question? It involves deciding whether to construct new measuring instruments or select instruments that have been developed by others. For instance, researchers might review existing multiple-choice tests and select those that are most valid for answering their research questions. They also might build or adapt existing interview schedules, questionnaires, personality scales, and so on, for use in making observations.

Timing and circumstances of the observations may affect the results.

4. They plan *when* the observations will be made, such as the time of day and/or duration of observations for the study. For instance, will the observations be made in the morning or late at night?

Researchers realize that the timing and circumstances of their observations may affect the results of their investigation.

5. They plan to make the observations *under particular circumstances*. For instance, will the observations be made in a quiet room or in a busy shopping mall? Will the observations be made in the presence of a treatment such as an experimental drug?[3]

Here's an example of an empirical research:

What?: What factors attribute to high or low academic achievement in a particular school district?

Whom?: High school-age children.

How?: Use school and district's information on student demographics, academic achievement scores, and conduct a survey for additional information that is not on the school database.

When?: Conduct the survey at the end of an academic year to be compiled along with the school database.

Under what circumstances?: Survey questionnaires will be distributed to be completed in class on the last week of the academic year with prior consent from the children's parents.

As illustrated in the sample, researchers plan *what, who, how, when,* and *under what circumstances* to observe.

Unfortunately, not all plans are good, and even the best plans often cannot be fully executed because of physical, ethical, legal, or financial constraints. Thus, empirical research varies in quality, and flawed research can be just as misleading as everyday observations often are.

Observations may be direct, such as watching adolescents in a multiracial group interact with each other, or they may be indirect, such as having adolescents respond to a questionnaire on how they would interact in an interracial setting.

Observations that researchers make result in data. Observations can yield numerical (quantitative) data or verbal (qualitative) data according to researchers' research questions.[4] An example of qualitative data might be descriptions of campus life experiences of first-generation college students. An example of numerical data might be scores on a scale that measures depression.

Large amounts of data need to be organized and summarized. The primary function of statistical analysis is to analyze quantitative data. For instance, a researcher could summarize quantitative data from an election poll by computing the percentage of those who plan to vote for each candidate, or a researcher could summarize depression scores by computing

Flawed research can be misleading.

Observations may be direct or indirect.

Observations result in data.

Large amounts of data need to be organized and summarized. The primary function of statistical analysis is to analyze quantitative data.

an average score. These and many other statistics are described in this book as a tool for understanding the world around us.

Notes on Terminology for Referring to Participants in Research

*The term **subjects** is appropriate when there is no consent.*

The term ***subjects*** is the traditional term for the individuals being studied. In recent decades, researchers have increasingly used the term ***participants***, which implies that the individuals being studied have freely consented to participate in the research. Note, however, that the term *subjects* is still appropriate when those being studied have not consented to participating. Examples include animals used in medical and psychological research, as well as individuals observed unobtrusively without their consent, such as adolescents observed in a shopping mall without their knowledge.

*The term **participants** is used when individuals have consented.*

Respondents refer to individuals who respond to a survey.

In addition to *subjects* and *participants*, other terms widely used by researchers are ***respondents***, which is most frequently used when individuals respond to a survey such as a political poll. And the term ***examinees*** is sometimes used in reference to participants who have taken an examination, such as an achievement test.

Examinees refer to those who take an examination.

Because the term *participants* is, by far, the predominant term used in research reports in the social and behavioral sciences, this term will be used throughout the book.

*The term **participants** is predominant.*

Exercise for Chapter 1

Factual Questions

1. The term *empiricism* refers to what?
2. Does everyday observation employ the empirical approach?
3. When researchers systematically use the empirical approach to acquire knowledge, we say that they are engaging in what?
4. What is the name of a subset of a population?
5. Which type of planning involves constructing or selecting measuring instruments (circle one)?
   A. Whom.
   B. How.
   C. When.
   D. Under what circumstances.

6. Even the best plans for research often cannot be fully executed for physical reasons. According to this chapter, what are some of the other reasons for this?
7. Observations that researchers make result in what?
8. Are the data that researchers collect always "scores"?
9. In recent decades, researchers have increasingly used what term to refer to the individuals being studied?

Questions for Discussion

10. Do you think that the opinions of your friends and acquaintances are good predictors of the outcomes of elections? Why? Why not?
11. Given some examples in this chapter, recall an instance in which you were misled by an everyday observation. Briefly describe it.
12. Recall an instance in which you read or heard about empirical research that you suspected was flawed. Briefly describe why you suspected that.

Notes

1. Other approaches are (1) *mathematical deduction*. This is when we deduce a proof in mathematics based on certain assumptions and definitions, and (2) *reliance on authority*, such as relying on a dictator's pronouncements as a source of knowledge.
2. Sampling methods are described in Part B.
3. When the researcher administers a treatment(s), the study is classified as an *experiment*. Experiments are discussed in Chapter 2.
4. Quantitative and qualitative types of studies are discussed in depth in Chapter 3 of this book.

CHAPTER 2

Types of Empirical Research

Chapter Objectives

The reader will be able to:

- ❏ Describe the distinction and purpose of non-experimental and experimental research.
- ❏ Identify the role of independent and dependent variables.
- ❏ Recognize the distinction between experimental and control groups.

A fundamental distinction within a quantitative study is whether the research is *non-experimental* or *experimental*. A ***non-experimental design***, which is sometimes called a *descriptive study*, can be descriptive in nature when little is known about a phenomenon.

In this design, observations are made to determine the status of what exists at a given point in time *without* administering treatments.

An example is a survey in which a researcher wants to determine participants' attitudes. In such a study, researchers strive *not* to change the participants' attitudes. They do this by avoiding leading questions and having the interviewers remain neutral in tone and mannerisms. If the sample is properly drawn and well-crafted questions are asked, a

A ***non-experimental design*** (descriptive study) can be descriptive in nature when little is known about a phenomenon. It is one in which no treatments are administered

DOI: 10.4324/9781003299356-3

researcher can obtain solid data on the attitudes held by participants. Note, however, that the researcher who has conducted such a non-experimental study has not gathered data on how to change attitudes. To do this, he or she would need to conduct an experiment.[1]

This involves descriptive questions of who, what, when, where, and how much or how many of various aspects of a phenomenon exist.

A non-experimental design can also be a correlational design that is exploratory and explanatory in nature, where there is some knowledge base to build on. Within this design, relationships between or among variables are proposed and the "why" and "how" questions of a phenomenon are asked. For example: what factors are related to students' success in school?

The predictive relationship can be tested, where the predictor is related to the outcome. However, these variables are variables that occur naturally and without any manipulations.

An ***experimental study*** is predictive in nature. Clearly, the purpose of an experimental study is to identify a ***cause-and-effect relationship***, in which treatments are given through a random assignment to see how the participants respond to the treatments.[2]

We all conduct informal experiments in our everyday lives that involve a treatment and a response. Here are some examples:

- ❏ We might try a new laundry detergent (the treatment) to see whether our clothes are cleaner (the response) than when we used our old brand.
- ❏ A teacher might bring to class a new short story (the treatment) to see whether students enjoy it (the response) or not.
- ❏ A waiter might try being more friendly (the treatment) to see whether it increases his tips (the response) or not.

The study tests for predictive questions or theories to address the effectiveness or a possible cause-and-effect, in which the independent variable is the possible cause and the dependent variable demonstrates the possible effect. When the hypothetical waiter mentioned earlier tries being friendlier, he is interested in finding out whether the increased friendliness (the independent variable) *causes* increased tips (the dependent variable).

In an experiment, the treatments are called the ***independent variable***, and the responses are called the ***dependent variable***. Independent variables are administered so that researchers can observe possible changes in dependent variables. For example: is there a change in the dependent variable per change in the independent variable?

A non-experimental design can be exploratory and explanatory in nature, where there is some knowledge basis to build on.

*An **experimental study** is predictive in nature.*

*The purpose of experiments is to determine **cause-and-effect relationships**.*

An experimental study is a design in which treatments are given to see how the participants respond to the treatments.

*Treatments are called the **independent variable** and responses are called the **dependent variable**.*

Unfortunately, informal experiments can be misleading. For instance, suppose a server at a restaurant notices that when he is friendlier, he gets larger tips than when he is less friendly. Did the increased friendliness *cause* the increase in the tips? The answer is not clear. Perhaps, by chance, the evening that the waiter tried being more friendly, he happened to have more generous customers. Perhaps an advice columnist published a column that day on tipping, which urged people to be more generous when tipping their server. Perhaps the server was not only more friendly but, unconsciously, also more efficient, and his increased efficiency, and not his increased friendliness, caused the increase in tips. The possible alternative explanations are almost endless unless an experiment is planned in advance to eliminate them.

> Unless it is properly planned, there may be many alternative explanations for the results of an experiment.

How could we conduct an experiment on friendliness and tipping that would allow for a clearer interpretation? The answer is by having an appropriate control condition. For instance, we could have the server be friendlier to every alternate party of customers. These customers would constitute the ***experimental group***, which receives a control condition. The remaining customers who receive the normal amount of friendliness would be referred to as the ***control group***. This is the group without a control condition. In addition, we could monitor the servers' behavior to be sure it is the same for both groups of customers in all respects, except for the degree of friendliness. Then, statistics could be used to compare the average tips earned under a more friendly condition with those earned under a less friendly condition.

> An appropriate control condition is an essential characteristic of good experiments.

> The **experimental group** receives a control condition.

> The **control group** receives no control condition.

Another type of empirical research worth mentioning, specifically when the field involves education or professional development is *program evaluation*. In the context of school-based or education-related study in any field, a ***program evaluation*** can be conducted to ascertain if a program in place for a specific group is effective. The goal of a program evaluation is to measure the quality, merit, and worth of the program's goals. A program evaluation may include a *needs assessment*, which measures the current issues and needs that shape the plans and goals of a program and its implementations and evaluative questions. This type of study may or may not be experimental in nature, but the distinction between program evaluation and research is that while research generally aims to generalize findings to a larger population, evaluation focuses on the effectiveness of a particular program's implementation for a specific group(s), whether it would be an intervention or a specialized curriculum, and so on. Evaluation can be conducted for a large group, but sometimes an evaluation can be conducted for a single case depending on the purpose

> ***Program evaluation*** measures the effectiveness of a program in place for a specific group.

of the evaluation. Program evaluation in the context of education helps to promote data-based decision making for education-related decisions.

Exercise for Chapter 2

Factual Questions

1. In which type of study are treatments given in order to see how participants respond?
2. In an experiment, are the responses an "independent variable" *or* a "dependent variable"?
3. What is the purpose of an experiment?
4. In an experiment, a researcher administered various dosage levels of aspirin to different groups of participants in order to determine the effects of various dosage levels on heart attack rates. In this study, "heart attack rates" is the (circle one):
   A. dependent variable.
   B. independent variable.
5. In an experiment, a researcher used group counseling with some participants and used individual counseling with other participants in order to study the effectiveness of the two types of counseling on raising the participants' self-esteem. In this study, the two types of counseling constitute the (circle one):
   A. dependent variable.
   B. independent variable.
6. In which type of study do researchers try *not* to change the participants?
7. What is the definition of a *non-experimental study*?
8. What is the distinction between an experimental group and a control group?
9. What is the main distinction between research and program evaluation?

Questions for Discussion

10. Briefly describe an informal experiment that you or someone you know recently conducted. Were there alternative explanations for the responses you observed?
11. Do you think that both *non-experimental* and *experimental* studies have a legitimate role in the acquisition of scientific knowledge? Why? Why not?

12. Give an example of a program evaluation report of a specific program that you may have conducted or read about.

Notes

1. For some important causal questions, it is not possible to conduct an experiment. For instance, in studying the effects of smoking and health, it would be unethical to encourage or force some human participants to smoke while forbidding others to do so. In such situations, researchers must use data collected in non-experimental studies to explore causality.
2. Random assignment means that each participant had an equal chance of being selected for either the control or experimental group. The concept of random assignment or random sampling will be discussed in Part B.

CHAPTER 3

Scales of Measurement

Chapter Objectives

The reader will be able to:

- ❏ Recall that variables are measurable factors that affect a phenomenon (or phenomena).
- ❏ Identify the different scales of measurement from the lowest to highest levels (nominal, ordinal, interval, ratio).

Variables refer to any measurable factor that has an effect on a phenomenon or phenomena. All variables must have at least two categories of possible answers with varying answers from the respondents. If you asked a group of people: "Are you alive and breathing?" while theoretically, a person could be alive or dead, the answers you collect will be "alive." And since the answers don't vary for this question, this variable does not help to explain a particular phenomenon, making this measurement not useful.

All variables can be categorized into four different *scales of measurement* (also known as *types of variables*). The scales of measurement help researchers determine what type of statistical analysis is appropriate for a given set of data. It is important to master the material in this chapter because it is referred to in a number of chapters that follow.

Variables are measurable factors that have an effect on a phenomenon.

All variables can be categorized into four different *scales of measurement*.

DOI: 10.4324/9781003299356-4

The lowest level of measurement is **nominal,** which is the *naming* level.

The lowest level of measurement is **nominal**. It is helpful to think of this level as the *naming* level because names (i.e., words) are used instead of numbers. Here are four examples:

- ❏ Participants name the political parties with which they are affiliated.
- ❏ Participants name their gender.
- ❏ Participants name the state in which they reside.
- ❏ Participants name their religious affiliation.

Notice that the categories that participants name in the above examples do not put the participants in any particular order. There is no basis on which we could all agree for saying that Republicans are either higher or lower than Democrats. The same is true for gender, state of residence, and religious affiliation.

Ordinal measurement puts participants in rank *order.*

The next level of measurement is **ordinal**. Ordinal measurement puts participants in rank *order* from low to high, but it does *not* indicate how much lower or higher one participant is in relation to another. To understand this level, consider these examples:

- ❏ Participants are ranked according to their height; the tallest participant is given a rank of 1, the next tallest is given a rank of 2, and so on.
- ❏ Three brands of hand lotion are ranked according to consumers' preferences for them.
- ❏ High school students rank order the subjects they are taking in school, giving their favorite subject a rank of 1, their next favorite a rank of 2, and so on.

Nominal and *ordinal* levels of measurements are known as **categorical data.**

In the examples, the measurements indicate the relative standings of participants but do not indicate the amount of difference among them. For instance, we know that a participant with a rank of 1 is taller than a participant with a rank of 2, but we do not know by how much. The first participant may be only one-quarter of an inch taller or may be two feet taller than the second. Nominal and ordinal levels of measurements are known as **categorical data**.

Interval and *ratio* scales measure *how much* participants differ from each other.

Interval and *ratio* scales are known as **numerical (or continuous) data**.

The next two levels, **interval** and **ratio**, tell us by *how much* participants differ. These scales are expressed in numbers, unlike categorical variables. Thus, interval and ratio scales are known as **numerical (or continuous) data**. For example:

- ❏ The height of each participant is measured to the nearest inch.
- ❏ An increase in students' math scores after receiving a math intervention course.

Notice that if one participant is 5 feet 6 inches tall and another is 5 feet 8 inches tall, we not only know the order of the participants in height, but we also know by how much the participants differ from each other (i.e., 2 inches). Both *interval* and *ratio* scales have equal intervals. For instance, the difference between 3 inches and 4 inches is the same as the difference between 5 inches and 6 inches.

In most statistical analyses, *interval* and *ratio* measurements are analyzed in the same way. However, there is a scientific difference between these two levels. An *interval* scale does not have an absolute (or non-arbitrary) zero. For instance, if we measure intelligence, we do not know exactly what constitutes absolute zero intelligence and thus cannot measure the zero point.[1] In contrast, a *ratio* scale has an absolute zero point on its scale.[2] For instance, we know where the zero point is on a tape measure when we measure height.

Ratio scales have an absolute (or non-arbitrary) zero; *interval* scales do not.

If you are having trouble mastering levels of measurement, memorize this environmentally friendly phrase: **N**o **O**il **I**n **R**ivers

No **O**il **I**n **R**ivers

The first letters of the words No Oil In Rivers (NOIR) are the first letters in the names of the four levels of measurement, in order from lowest to highest complexity of data. Now read this chapter again and associate the definitions with each level. Table 3.1 summarizes the scales of measurements with examples.

TABLE 3.1
Descriptions of Four Types of Variable

Description	Categorical Data		Continuous Data	
	Nominal	Ordinal	Interval	Ratio
Categories	✓	✓	✓	✓
Order (rank)		✓	✓	✓
Equal intervals			✓	✓
Absolute zero				✓
Example:	Political parties	Favorite subject	Intelligence	Height

Variables can be expressed in various scales of measurement depending on the purpose of the analysis and research question. For example, take "years of formal education." The study may be focused on the distinction between primary and secondary education, in which case we can transform the numbers of the variable into the two categories. This would transform the data to a nominal data. This type of variable that has two categories that are mutually exclusive between

A variable that has two categories that are mutually exclusive between the two categories, but inclusive of all participants in the two categories, is known as a ***dichotomous variable.***

A variable can be transformed into a different scale depending on the purpose of the study.

the two categories, but inclusive of all participants in the two categories, is known as a ***dichotomous variable***.

But if the study is interested in the "elementary, middle, and high school" categories in relation to their curricular challenges, for example, the numbers can be transformed into these three categories which would become an ordinal data. Or the study can list the number of years of formal education in an educational setting to be a ratio data to predict its effect on their children's education attainment, for example. So, a variable can be transformed into a different scale depending on the purpose of the study.

Understanding the different types of variables is critical in choosing the right type of analysis. Table 3.2 illustrates this point. While the terminologies in the table have not yet been discussed, one can see that variables are analyzed differently based on their type. For example, for a nominal data, one would report the mode (the most frequently occurring category) as the appropriate measure of central tendency instead of median or mean.

The table is a summary of descriptive statistics. Thus, using this table as an outline, an in-depth explanation of various descriptive statistics will be discussed in subsequent chapters. But first, an overview of descriptive, correlational, and inferential statistics will be discussed in the next chapter.

TABLE 3.2

An Illustration of Various Descriptive Statistics Based on Types of Variable

Types of Variable	Distribution	Measures of Central Tendency	Measures of Variability	Correlations (Relationship)
Nominal	Frequency distribution	Mode	N/A	*phi*
Ordinal	Frequency distribution	Mode/ median	N/A	Spearman *r*
Interval, Ratio (skewed)	Frequency distribution	Median	Range, (semi) interquartile range	None
Interval, Ratio (normal)	Frequency distribution	Mean	Variance, standard deviation	Pearson *r*

Exercise for Chapter 3

Factual Questions

1. What is the name of the lowest scale of measurement?
2. Which level of measurement should be thought of as the "naming" level?
3. Which scale of measurement puts participants in rank order?
4. Which two scales of measurement indicate the amount by which participants differ from each other?
5. Which scale of measurement has an absolute zero?
6. If you measure the weight of participants in pounds, which scale of measurement are you using?
7. If you rank employees from most cooperative to least cooperative, which scale of measurement are you using?
8. If you ask participants to name the country they were born in, which scale of measurement are you using?
9. What phrase should you memorize in order to remember the scales of measurement in order?
10. Which scale of measurement is between the ordinal and ratio scales?

Question for Discussion

11. Name an example of each scale of measurement: nominal, ordinal, interval, and ratio.

Notes

1. Most applied researchers treat the scores from standardized tests (except percentile ranks and grade-equivalent scores) as *interval* scales of measurement.
2. Thus, the ratio scale is the only one for which it is appropriate to compute ratios. For instance, it is appropriate to make statements such as "John is twice as tall as Sam" (a ratio of 2 to 1) only when using a ratio scale.

CHAPTER 4

Descriptive, Correlational, and Inferential Statistics

Chapter Objectives

The reader will be able to:

- ❏ Identify the role of descriptive, correlational, and inferential statistics.
- ❏ Recall that descriptive statistics summarizes the data.
- ❏ Recall that correlational statistics describes the relationship of two or more variables of data.
- ❏ Recall that inferential statistics is used to generalize results of a sample to a population.

Quantitative studies use statistics to analyze data based on the type of data, as discussed in Chapter 3. Statistics largely fall into descriptive and inferential statistics. ***Descriptive statistics*** summarizes the data, providing descriptions of certain characteristics of a given data. For instance, suppose you have the scores on a standardized test for 500 participants. One way to summarize the data is to calculate an ***average*** score, which is a descriptive statistic that indicates the scores overall. You might also determine the ***range*** of scores from the highest to the lowest, which would indicate how much the scores vary

Descriptive statistics summarize data.

An **average** is a descriptive statistic.

A **range** of scores is between the highest and the lowest scores.

DOI: 10.4324/9781003299356-5

from the highest to the lowest score. These and other descriptive statistics are described in detail in Part C.

Correlational statistics, which are described separately in Part D of this book, are a special subgroup of descriptive statistics. The purpose of *correlational statistics* is to describe the relationship between two or more variables for a group of participants. For instance, suppose a researcher is interested in the predictive validity of a college admissions test. The researcher could collect the admissions test scores and the freshman GPAs for a group of college students. To determine the relationship between the admissions test scores and GPAs or the validity of the test for predicting GPAs, a statistic known as a *correlation coefficient* can be computed. Correlation coefficients range in value from 0.00 (no correlation between variables) to 1.00 or −1.00 (a perfect correlation).[1] For example, shoe size and intelligence have no correlation (correlation coefficient is 0). However, mothers' height and daughters' height have a strong correlation (correlation coefficient would be closer to 1.00). The strength of a relationship, as well as the direction of a relationship (positive, negative), are discussed in detail in Part D.

Inferential statistics takes descriptive statistics further by extending the results from a sample to the population that the sample represents. Thus, inferential statistics are tools that tell us how much confidence we can have when generalizing results from a sample to a population.[2] Consider national opinion polls in which carefully drawn samples of only about 1,500 adults are used to estimate the opinions of the entire adult population of the United States. The pollster first calculates *descriptive statistics*, such as the *percentage* of respondents in favor of capital punishment and the percentage opposed.

Inferential statistics, as the name implies, makes inferences about a larger population from a sample of that population. However, having sampled, the researcher knows that the results may not be accurate because the sample may not be representative. In fact, the pollster knows that there is a high probability that the results are off by at least a small amount. This is why pollsters often mention a *margin of error*, which is an inferential statistic that measures to what degree the results from the sample are inaccurate as caused by the sample representing the population. It is reported as a warning to readers of research that random sampling may have produced errors, which should be considered when interpreting results. For instance, a weekly news magazine recently reported that 52% of the respondents in a national poll believed the economy was improving. A footnote in the report indicated that the margin of error was ±2.3%. This means that the pollster was confident that

---

The purpose of **correlational statistics** is to describe the relationship between two or more variables for one group of participants.

Correlation coefficients range in value from 0.00 to 1.00 or −1.00.

Inferential statistics are tools that tell us how much confidence we can have when generalizing from a sample to a population.

A **percentage** is a descriptive statistic.

A **margin of error** is an inferential statistic.

the true percentage for the whole population was within 2.3% percentage points around 52% (i.e., 49.7%–54.3%).[3]

You may recall from Chapter 3 that a *population* is any group in which a researcher is interested in studying. It may be large, such as all US residents over age 18, or it may be small, such as all registered nurses employed by a specific hospital. A study in which all members of a population are included is called a ***census***. A census is often feasible and desirable when studying small populations (e.g., an algebra teacher may choose to pretest all students at the beginning of a course). When a population is large, it is more economical to study only a sample of the population. With modern sampling techniques, highly accurate information can be obtained through the use of relatively small samples. For this reason, sampling and variations of random sampling, sample size, and standard error of the mean are important topics that are discussed in detail in Part B.

Inferential statistics are *not* needed when analyzing the results of census data since this would be a study of all people, and there is no sampling involved. The use of various inferential statistics for evaluating results involving sampling is discussed in Part E.

A ***census*** is a study in which all members of a population are included.

Inferential statistics are *not* needed when analyzing the results of a census.

Exercise for Chapter 4

Factual Questions

1. Is an *average* a "descriptive statistic" *or* an "inferential statistic"?
2. Is the *range* of a set of scores a "descriptive statistic" *or* an "inferential statistic"?
3. What is the purpose of correlational statistics?
4. If there is no relationship between two sets of scores, what is the value of the correlation coefficient?
5. Inferential statistics are tools that tell us what?
6. Is a margin of error a "descriptive statistic" *or* an "inferential statistic"?
7. A margin of error is reported as a warning to readers about what might have happened in relation to sampling?
8. What is a study in which all members of the population are included?
9. Why are inferential statistics *not* needed when analyzing the results of a census?

Questions for Discussion

10. Give an example of a study that would require the use of descriptive statistics.
11. Give an example of a study that would require the use of a correlational study.
12. Give an example of a study that would require the use of inferential statistics.

Notes

1. Correlation coefficients can also be negative, indicated by a negative correlation coefficient. This topic and the interpretation of correlation coefficients are discussed in Part D.
2. The word *inferential* comes from *infer*. When we generalize from a sample to a population, we are *inferring* that what's true about the sample is also true for the representing population.
3. 52% − 2.3% = 49.7%, 52% + 2.3% = 54.3%.

PART B

SAMPLING

In the discussion of descriptive and inferential statistics, an important topic to consider is sampling. Oftentimes, descriptive statistics describes a set of data in a sample, and when the sample represents the population, then one can make inferences about the population from the sample. The techniques of sampling and the concept of biased and unbiased sampling methods and sampling error, as well as the benefits of unbiased random sampling, are discussed in Part B before delving further into a detailed conceptual discussions of descriptive and inferential statistics. The concept of the standard error of the mean and central limit theorem are also discussed in this section.

Chapter 5. Introduction to Sampling
Chapter 6. Random Sampling
Chapter 7. Sample Size
Chapter 8. Standard Error of the Mean and Central Limit Theorem

DOI: 10.4324/9781003299356-6

Introduction to Sampling

Chapter Objectives

The reader will be able to:

- ❏ Explain various techniques of sampling.
- ❏ Explain the difference between biased and unbiased sampling.
- ❏ Recognize that sampling (both unbiased and biased) creates sampling error.
- ❏ Evaluate the benefits of unbiased random sampling.
- ❏ Demonstrate how a simple random sampling is conducted.

As defined in the Introduction, a ***population*** consists of all members of a group in which a researcher has an interest. It may be small, such as all psychiatrists affiliated with a particular hospital, or it may be large, such as all high school seniors in a state. When populations are large, researchers usually use a sample of the population. So, a ***sample*** is a subset of a population. For instance, we might be interested in the attitudes of all registered nurses in Texas toward people with AIDS. The nurses would constitute the population. If we administered an AIDS attitude scale to all these nurses, we would be studying the population, and the summarized results (such as averages)

A ***population*** consists of all members of a group.

A ***sample*** is a subset of a population.

DOI: 10.4324/9781003299356-7

would be referred to as **parameters**. If we had studied only a sample of the nurses, the summarized results would be referred to as **statistics**.

No matter how a sample is drawn, it is always possible that the *statistics* obtained by studying the sample do not accurately reflect the *population parameters* that would have been obtained if the entire population had been studied. In fact, researchers almost always expect some amount of error as a result of sampling.

If sampling creates errors, why do researchers sample? First, for economic and physical reasons, it is not always possible to study an entire population. Second, with proper sampling, highly reliable results can be obtained. Furthermore, with proper sampling, the amount of error to allow for in the interpretation of the resulting data can be estimated with inferential statistics, which are covered in Part E of this book.

Freedom from **bias** is the most important characteristic of a good sample. Bias exists whenever some members of a population have a greater chance of being selected for inclusion in a sample than other members of the population, creating a **biased sample**. Here are three examples of biased samples:

- ❏ A professor wishes to study the attitudes of all sophomores at a college (the population) but asks only those enrolled in her introductory psychology class (the sample) to participate in the study. Note that only those in the class have a chance of being selected; other sophomores have no chance.
- ❏ An individual wants to predict the results of a citywide election (the population) but asks about the intentions of voters who he only encounters in a large shopping mall (the sample). Note that only those in the mall have a chance of being selected; other voters have no chance.
- ❏ A magazine editor wants to determine the opinions of all rifle owners (the population) on a gun-control measure but mails questionnaires only to those who subscribe to her magazine (the sample). Note that only subscribers to her magazine have a chance to respond; other rifle owners have no chance.

In the above examples, **samples of convenience** (or **accidental samples**) were used, increasing the odds that some members of a population would be selected while reducing the odds that other members would be selected. In addition to the obvious bias in the examples, there is an additional problem. Even those who do have a chance of being included in the samples may refuse to participate. This problem is often referred to as the problem of **volunteerism** (also called *self-selection bias*). Volunteerism is presumed to create an additional source of bias because those who decide not to participate have no chance of being included. Furthermore, many studies comparing participants (i.e., volunteers) with non-participants suggest that participants tend to be more highly educated and tend to come from higher socioeconomic status (SES) groups than their counterparts. Efforts to reduce the effects of volunteerism include offering rewards, stressing to potential participants the importance of the study, and making it easy for individuals to respond, such as by providing them with a stamped, self-addressed envelope if responding by mail.

To eliminate bias in the selection of individuals for a study, some type of *random sampling* is needed. A classic type of random sampling is *simple random sampling*. This technique gives each member of a population an equal chance of being selected. A simple way to accomplish this with a small population is to put the names of all members of a population on slips of paper, thoroughly mix the slips, and have a blindfolded assistant select the number of slips desired for the sample.[1] After the names have been selected, efforts must be made to encourage all those selected to participate. If some refuse, as often happens, a biased sample is obtained even though all members of the population had an equal chance to have their names selected.

Suppose that a researcher is fortunate. The researcher selected names using simple random sampling and obtained the cooperation of everyone selected. In this case, the researcher has obtained an ***unbiased sample***, where individuals had an equal chance of being selected. Can the researcher be certain that the results obtained from the sample accurately reflect those results that would have been obtained by studying the entire population? Certainly not. The possibility of random errors still exists. Random errors (created by random selection) are called ***sampling errors*** by statisticians. Sampling errors measure the degree of misrepresentation of the sample from the population. At random (i.e., by chance), the researcher may have selected a disproportionately large number of Democrats, males, low SES group members, and so on. Such errors make the sample unrepresentative, and therefore may lead to incorrect results. Variations of sampling will be discussed in greater depth in Chapter 6.

If both biased and unbiased sampling are subject to error, why do researchers prefer unbiased random sampling? They prefer it for two reasons: (1) inferential statistics, which are described in Part E of this book, enable researchers to estimate the amount of error to allow for when analyzing the results from unbiased samples, and (2) the amount of sampling error obtained from unbiased samples tends to be small when large samples are used.

Random sampling yields an unbiased sample.

In *simple random sampling*, each member of a population is given an equal chance of being selected.

Simple random sampling identifies an *unbiased sample*.

Sampling errors measure the degree of misrepresentation of the sample from the population.

Random sampling produces sampling errors.

While using large samples helps to reduce the amount of random error, it is important to note that selecting a large sample does not correct for errors due to bias. For instance, if the individual, trying to predict the results of a citywide election in the second example above, is very persistent and spends weeks at the shopping mall asking shoppers how they intend to vote, the individual will obtain a very large sample of people who may differ from the population of voters in various ways, such as by being more affluent, having more time to spend shopping, and so on. This illustrates that increasing the size of a sample does not reduce the amount of error due to bias.

> Selecting a large sample does not correct for errors due to bias.

Despite the above discussion, it is *not* true that all research in which biased samples are used is worthless. There are many situations in which researchers have no choice but to use biased samples. For instance, for ethical and legal reasons, some medical research is conducted using volunteers who are willing to risk taking a new medication or undergoing a new surgical procedure. If promising results are obtained in initial studies, larger studies with better (but usually still biased) samples are undertaken. At some point, despite the possible role of bias, decisions, such as Food and Drug Administration approval of a new drug, need to be made on the basis of data obtained with biased samples. Little progress would be made in most fields if the results of all studies with biased samples were summarily dismissed. However, when biased samples are used, the results of statistical analyses of the data should be viewed with great caution.

> Often, researchers have no choice but to use biased samples.

At the same time, it is important to note that the statistical remedies for errors due to biased samples are extremely limited. Because researchers usually do not know the extent to which a particular bias has affected their results (e.g., they do not know how non-respondents to a questionnaire would have answered the questions), it is generally not possible to adjust statistically for errors created by bias.

> Statistical results based on observations of biased samples should be viewed with great caution.

Various methods of random sampling are described in more detail in Chapter 6. Considerations in determining sample size are discussed in Chapter 7.

Exercise for Chapter 5

Factual Questions

1. What term is used to refer to all members of a group in which a researcher has an interest?
2. If samples yield "statistics," what do populations yield?

3. What is the most important characteristic of a good sample?
4. If a researcher uses a sample of volunteers from a population, should we presume that the sample is biased?
5. What type of sampling eliminates bias in the selection process of participants?
6. Briefly describe how one could select a simple random sample.
7. Does random sampling produce sampling errors?
8. The amount of random sampling error obtained from unbiased samples tends to be small when what is done?
9. Is selecting a large sample an effective way to reduce the effects of bias in sampling? Explain.
10. According to this chapter, is research in which the sample is biased worthless?

Questions for Discussion

11. Are you convinced that using a rather small, unbiased sample is better than using a very large, biased sample? Why? Why not?
12. Be on the lookout for a news report of a scientific study in which a biased sample was used. If you find one, briefly describe it.

Note

1. Another method for selecting a *simple random sample*, and other types of random samples, is described in Chapter 6.

CHAPTER 6

Random Sampling

Chapter Objectives

The reader will be able to:

❑ Differentiate simple random, stratified random, and random cluster sampling.
❑ Describe the use of a table of random numbers.
❑ Recognize that while there are sampling methods to minimize sampling error, random sampling is not without sampling error.

As noted in Chapter 1, a *population* consists of all members of the group the researcher is interested in studying. A population may be small, such as all social workers employed by a public hospital in Detroit, or it may be large, such as all social workers in Michigan. The larger the population, the more likely the researcher will study only a sample of the population and *infer* that what is true for the sample is also true for the population. The process of making such an inference is referred to by statisticians as *generalizing* from a sample to a population.

Freedom from bias is the most important characteristic of a sample.[1] An *unbiased sample* is defined as one in which all individuals in a population have an equal chance of being included as a participant. The basic

DOI: 10.4324/9781003299356-8

method for obtaining an unbiased sample is to use *random sampling* from a population. Various types of random sampling are described below.

To draw a ***simple random sample***, a researcher can put names on slips of paper and draw the number needed for the sample. This method is efficient for drawing samples from small populations.

To draw a simple random sample from a large population, it is more efficient to use a ***table of random numbers*** than to write names on slips of paper. A portion of a table of random numbers is shown in Appendix C at the end of this book,[2] and the first two rows of the table are shown in Table 6.1.

In this table, there is no sequence to the numbers, and in a large table of random numbers, each number appears about the same amount of times. To use the table, first assign everyone in the population a *number name*. For instance, if there are 90 individuals in a population, name the first individual 01, the second individual 02, the third individual 03, and so on, until you reach the last individual, whose number is 90.[3] (Often, computerized records have the individuals already numbered, which simplifies the process. Any set of numbers will work as number names as long as each individual has a different number and all individuals have the same number of digits in their number names.) To use the table, flip to any page in a book of random numbers and put your finger on the page without looking. This will determine the starting point.

Let us start in the upper left-hand corner of Table 6.1 for the sake of illustration. Because each individual has a two-digit number name, the first two digits identify the first participant; this is individual number 21. The next two digits to the right are 0 and 4; thus, individual number 04 will also be included in the sample. The third number is 98. Because there are only 90 in the population, skip 98 and continue to the right to 08, which is the number of the next individual drawn. Continue moving across the rows to select the sample.

Stratified random sampling is useful for a larger population, whereas simple random sampling draws samples from small populations. In *stratified random sampling*, participants are drawn at random, separately from each stratum. The population is first divided into strata that are believed

TABLE 6.1
First Two Rows of Table of Random Numbers

Row #																			
1	2	1	0	4	9	8	0	8	8	8	0	6	9	2	4	8	2	6
2	0	7	3	0	2	9	4	8	2	7	8	9	8	9	2	9	7	1

to be relevant to the variable(s) being studied. Suppose, for instance, you wanted to survey opinions on alcohol consumption among all students on a college campus. If you suspect that freshmen and seniors might differ in their opinions on alcohol consumption, it would be desirable to first stratify the population according to years in college and then draw separately from each stratum at random. Specifically, you would draw a random sample of freshmen and then separately draw a random sample of seniors. The same percentage should be drawn from each stratum. For instance, if you want to sample 10% of the population and there are 2,000 freshmen and 1,600 seniors, you would draw 200 freshmen and 160 seniors. Notice that there are more freshmen in the sample than seniors, but this is appropriate because the freshmen consist of a larger group in the population. It is important to note that you would *not* be stratifying in order to compare freshmen with seniors. Rather, the purpose of stratifying is to obtain a single sample of the college population that is representative in terms of years in college.[4] This is also known as ***proportionate stratified sampling***, in which the same proportions of subgroups in the population are retained in the sample. In a historical research context, for example, the census demographic information can be used to stratify a community of interest, then use a simple random or systematic sampling method to generate a sample.

Proportionate stratified random sampling has the advantage of having appropriate proportions of freshmen and seniors in the sample. If your hunch was correct that freshmen and seniors differ in their opinions on alcohol consumption, you would have increased the precision of the results by stratifying.[5]

Note that stratifying does not eliminate all sampling errors. For instance, when drawing the seniors at random, you may, by chance, obtain seniors for the sample who are not representative of all seniors on campus; the same, of course, holds true for freshmen. Nevertheless, stratifying would eliminate the possibility of obtaining a disproportionately large number of either freshmen or seniors for the sample (note that it eliminates a particular *type* of sampling error and *not* all sampling errors).

For even larger-scale studies, ***multistage random sampling*** may be used (Figure 6.1). In this technique, a researcher might do the following. Stage 1: draw a sample of counties at random from all counties in a state. Stage 2: draw a sample of voting precincts at random from all precincts in the counties previously selected. Stage 3: draw individual voters at random from all precincts that were sampled. In multistage sampling, the researcher could introduce stratification. For instance, the researcher could first stratify the counties by the county size and then separately

Draw the same percentage, not the same number, from each stratum.

In **proportionate stratified sampling**, the same proportions of subgroups in the population are retained in the sample.

Multistage random sampling may be used in large-scale studies.

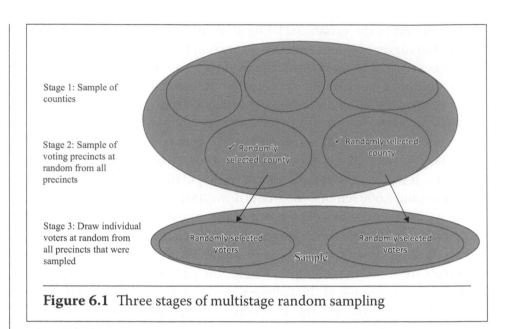

Stage 1: Sample of counties

Stage 2: Sample of voting precincts at random from all precincts

Stage 3: Draw individual voters at random from all precincts that were sampled

✓ Randomly selected county

✓ Randomly selected county

Randomly selected voters

Randomly selected voters

Sample

Figure 6.1 Three stages of multistage random sampling

In *random cluster sampling*, existing groups of participants are drawn.

draw counties at random by the various sizes of the counties, thereby ensuring that county sizes are taken into the sampling process.

Another technique that is sometimes useful is *random cluster sampling*. To use random cluster sampling, all members of a population must belong to a cluster (i.e., an existing group). For example, all Boy Scouts belong to troops; in most high schools, all students belong to homerooms; and so on. Unlike simple random sampling, in which individuals are drawn, in cluster sampling, *clusters* are drawn. To conduct a survey of Boy Scouts, for instance, a researcher could draw a random sample of troops, contact the leaders of the selected troops, and ask them to administer the questionnaires to all members of their troops.

There are some advantages and disadvantages to *random cluster sampling* (Figure 6.2). The two advantages to random cluster sampling over *sample random*, *stratified*, and *multistage random sampling* are as follows. (1) There are fewer individuals for the researcher to contact (e.g., only the troops' leaders and not the individual scouts), and for this reason, this sampling method is useful for studying a large geographical area. (2) The degree of cooperation is likely to be greater if troop leaders ask the individual scouts to participate than if a researcher who is unknown to the scouts asks them.

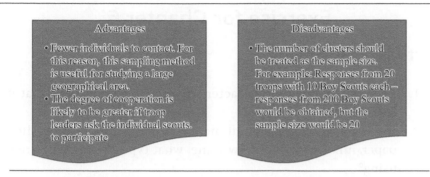

Figure 6.2 Advantages and disadvantages of random cluster sampling

TABLE 6.2
A Summary Table of Types of Random Sampling

Types of Random Sampling	Description
Simple random sampling	Draw names. Efficient for drawing samples from a small population.
Stratified random sampling	Divide the population into strata. Useful for a larger population.
Multistate random sampling	Involves three stages of dividing the population. Useful for even larger populations.
Random cluster sampling	Divide the population into existing groups called clusters first. This method accommodates large geographical studies.

However, there is also a disadvantage to random cluster sampling. For statistical reasons that are beyond the scope of this book, the number of clusters (not the number of participants) should be treated as the sample size. Thus, for instance, if 20 troops with 10 Boy Scouts each were selected, responses from 200 Boy Scouts would be obtained. However, the researcher would need to report the sample size as 20, not as 200. As a result, when using cluster sampling, it is desirable to use a large number of clusters to overcome the disadvantage.

Table 6.2 is a summary of various techniques of random sampling discussed in this chapter.

Exercise for Chapter 6

Factual Questions

1. The most important characteristic of a good sample is that it is free from what?
2. If you put the names of all members of a population on slips of paper, mix them, and draw some, what type of sampling are you using?
3. If there are 60 members of a population and you give them all number names starting with 01, what are the number names of the *first two participants selected* if you select a sample starting at the beginning of the third row of the Table of Random Numbers in Appendix C near the end of this book?
4. If there are 500 members of a population and you give them all number names starting with 001, what are the number names of the *first two participants selected* if you select a sample starting at the beginning of the fourth row of the Table of Random Numbers in Appendix C near the end of this book?
5. In what type of sampling is the population first divided into strata that are believed to be relevant to the variable(s) being studied?
6. Suppose you draw at random the names of 5% of the registered voters separately from each county in a state. What type of sampling are you using?
7. Does stratification eliminate all sampling errors?
8. Suppose you draw a sample of 12 of the homerooms in a school district at random and administer a questionnaire to all students in the selected homerooms. What type of sampling are you using?
9. Suppose you draw a random sample of 20 hospitals from the population of hospitals in the United States, then draw a random sample of maternity wards from the 20 hospitals, and then draw a random sample of patients in the maternity wards previously selected. What type of sampling are you using?

Question for Discussion

10. Suppose you want to conduct a survey of a sample of the students registered at your college or university. Briefly describe how you would select the sample.

Notes

1. Examples of biased samples are provided in Chapter 3.
2. Academic libraries have books of random numbers. Statistical computer programs can also generate them.
3. The number of digits in the number names must equal the number of digits in the population total. For instance, if there are 500 people in a population, there are three digits in the total and there must be three digits in each number name. Thus, the first case in the population is named 001, the second one is named 002, and so on.
4. If your purpose was to compare men with women, then it would be acceptable to draw the same number of each and compare averages or percentages for the two samples.
5. Of course, if your hunch that men and women differ in their opinions was wrong, the use of stratification would be of no benefit.

CHAPTER 7

Sample Size

Chapter Objectives

The reader will be able to:

- ❏ Recognize that deciding on a sample size is a complex issue.
- ❏ Recognize that a larger sample size does not always yield precision.
- ❏ Explain that a sample size should be based on variations in the population and expected findings in order to achieve precision.

In Chapter 5, you learned the importance of using random samples whenever possible. By using random sampling, researchers draw unbiased samples. However, an unbiased random sample still contains random *sampling error*. In other words, by the luck of the random draw, a random sample may differ from a population in important aspects. Fortunately, sampling error can be evaluated with inferential statistics, which is the topic of the remaining chapters in Part B. It is important to note, however, that inferential statistics cannot be used to evaluate the role of bias, which is why it is important to eliminate bias in the first place by using random sampling.

As a general rule, the larger the random sample, the smaller the sampling error. In technical terms, the larger the random sample is, the more *precise* the results are. Statisticians define *precision* as the extent to

An unbiased random sample contains *sampling error*.

The larger the sample, the smaller the sampling error and the greater the *precision*.

Precision is the extent to which the same results would be obtained with another random sample from the same population.

DOI: 10.4324/9781003299356-9

which the same results would be obtained if another random sample were drawn from the same population. The basic way to increase precision is to increase the sample size.[1] For instance, if a researcher drew two random samples of 500 individuals from a population, the two sets of results (e.g., the means for each sample of 500) would probably be closer than if the researcher had drawn two random samples of only 25 individuals each. Thus, a sample of 500 would be expected to have greater precision than a sample of 25.

While it is rather obvious that increasing sample size increases precision, it is less obvious that an increase in sample size to an already larger sample size would give the researcher ***diminishing returns.*** To understand diminishing returns, consider Examples 1 and 2.

> An increase in sample size to an already larger sample size gives the researcher ***diminishing returns*** in terms of precision.

EXAMPLE 1
Researcher A plans to sample ten individuals at random from a large population. At the last minute, the researcher decides to increase precision by increasing the sample size to 20 by selecting ten additional individuals for the sample.

EXAMPLE 2
Researcher B plans to sample 1,000 individuals at random from a large population. At the last minute, the researcher decides to increase precision by increasing the sample size to 1,010 by selecting ten additional individuals for the sample.

In both examples, the researchers have added ten individuals to their sample. Which researcher gets a bigger increase in precision by increasing the sample size by ten? The answer is Researcher A in Example 1. Researcher A has doubled the sample size from ten to 20 (a 100% increase), while Researcher B in Example 2 has increased their sample by only 1% (from 1,000 to 1,010).

Looking at Examples 1 and 2 from another perspective, we can see that the researcher in Example 1 could get a substantial change in results by increasing the sample size from ten to 20. In contrast, the researcher in

Example 2 can expect little effect on the results by adding ten to a sample of 1,000, giving the research diminished returns.

Because of the principle of diminishing returns, even when researchers are conducting national opinion polls, samples of about 1,500 are usually quite adequate. Typically, adding more individuals to a sample of this size produces very little increase in the precision of results.

Researchers often explore differences among groups of individuals. A general principle in determining sample size for examining group differences is this: *the smaller the anticipated difference in the population, the larger the sample size should be.* For instance, suppose a new drug has been developed for relieving headaches and the pharmaceutical researchers anticipate that the new drug will be only slightly more effective than existing drugs. With very small samples (one sample taking the new drug and one sample taking existing drugs), researchers may not be able to detect the small difference because the results will be imprecise due to the limited sample size.

A corollary is this: *even small samples can identify very large differences.* Suppose a researcher is testing a new antiviral treatment for a new strain of virus. The researcher uses random samples of 100 for the experimental group and 100 for the control group. (All individuals in both groups recently became ill with the virus.) Suppose, further, that of those who received the new treatment (the experimental group), only five died (5%), while of those who did not receive the new treatment, 45 died (45%). Because of the large difference, this should be regarded as a very promising result even though the researcher studied a total of only 200 individuals. To put this into perspective, consider this practical question: if you became ill with the virus, would you take the new treatment even though there were only 100 in the treatment group (a small sample for this type of study), *or* would you reject the treatment and wait until the result was later replicated in studies with larger samples? If you decided to take the treatment, you would be acknowledging the importance of the large difference obtained with the two relatively small samples.

When researchers determine the sample size for their studies, they should also consider the variation in the population because of this principle: *for populations with very limited variability, even small samples can yield precise results.* For instance, if you take a random sample of eggs that have already been graded as "extra-large" and weigh each egg, you will probably find only a small amount of variation among them.[2] For this population, a small random sample should yield a precise estimate of the average weight of "extra-large" eggs. The principle is made clearer by considering the highly hypothetical situation in Example 3: what size sample should the anthropologist use? Because there is essentially no variation in heights among identical twins who were raised identically, a sample size of one (one tribe member) should yield a highly precise answer.

For national opinion polls, samples of about 1,500 are usually adequate.

The smaller the anticipated difference in the population, the larger the sample size should be.

Even small samples can identify very large differences.

For populations with very limited variability, even small samples can yield precise results.

A corollary is this: *the more variable the population, the larger the sample size should be.* For instance, suppose you wanted to estimate the math achievement of sixth graders in a very large metropolitan school district and drew a random sample of only 30 students. Because there is likely to be tremendous variation in math ability across a large school district, a sample of 30, even though it is random, might yield very misleading results because of the lack of precision. By chance, for instance, you might obtain a disproportionately large number of high achievers. Using a much larger sample would greatly increase precision and thus reduce such a possibility.

Another principle in determining sample size is this: *when studying a rare phenomenon, large samples are usually required.* For example, suppose you wanted to estimate the percentage of college students who are HIV-positive. If you draw a sample of only about 100, you probably would find no cases because the disease is relatively rare and is unlikely to be evident in such a small sample; thus, you might mistakenly conclude that no college students are HIV-positive. By using a sample of thousands, you could get a more precise estimate of the small percentage that is positive.

EXAMPLE 3

An anthropologist discovers a lost tribe that consists of 1,000 identical twins. They are not only genetically identical, but they were also all raised in the same way, ate the same diets, and so on. The anthropologist wants to estimate the average heights of the members of the tribe by using a sample.

The most important principle to remember when considering sample size is this: *using a large sample does not correct for bias.* For instance, if you are homeless and ask hundreds of your homeless friends how they feel about the government giving a $5,000 (in US dollars) grant to each homeless individual, you may misjudge the opinions of the general public, which is your population. Even if you traveled around the country and asked thousands of homeless people you encountered about this issue, you probably would be just as much in error as you would be with a smaller, biased sample of just your homeless friends. This example illustrates that, in general, it is better to use a small, unbiased sample than a large, biased sample. Using a random process to select participants eliminates sampling bias.

So, what sample size is enough if the researcher wants to be sure about their results? Generally, when considering a research study, statistical power helps determine the sample size you need in order to get a statistically significant result. **Power analysis** helps determine what sample size we need to correctly conclude the significant difference. It is generally accepted that a statistical power of 0.8 (from a range of 0 ~ 1) is a strong enough power. This means that you have an 80% chance of detecting an effect if the actually exists.

So, in order to determine the sample size needed for a study, three components are required: the statistical power of choice (at least 0.8), the significance level (commonly alpha of 5%, which indicates that we allow a 5% chance of error in our study), and finally the effect size (d).

One way to calculate the effect size between two means is:

> **Effect size (d)** = the estimated difference between two means divided by pooled estimated standard deviation. In other words, the mean difference is divided by the average of the two standard deviations. *(See Chapter 32 for the computation formula of Cohen's d.)*

The statistical power of 0.8, alpha level of 0.5, and the effect size (calculated using means and standard deviations) are then plugged into a statistics power calculator or an online calculator.

Depending on what online or software calculator you use, you will need different components. For example, we can use IBM SPSS to estimate the sample size needed by estimating the two means and the standard deviation (perhaps from a pilot study) at a statistical power and alpha level of the researcher's choice.

If power analysis indicates a sample size of 100, for example, this means that if you have a sample size of 100, you have an 80% chance of being able to correctly conclude a significant difference (correctly reject the null hypothesis).

Concluding Comment

As you can see from the discussion above, determining sample size is a complex process. Thus, there is no simple answer to the question of how large a sample should be. As you work through the remainder of this book,

---

Power analysis helps determine what sample size we need to correctly conclude the significant difference

Effect size (d) = the estimated difference between two means divided by pooled estimated standard deviation.

sample size will be discussed again in various contexts because sample size affects the results obtained with inferential statistics, which is the topic of the remaining chapters in Part B.

Exercise for Chapter 7

Factual Questions

1. How do statisticians define the term *precision*?
2. What is the basic way to increase precision?
3. Suppose that Researcher Doe increased her sample size from 100 to 120, while Researcher Smith increased his sample size from 500 to 520. Which researcher will get a greater increase in precision by increasing the sample size by 20?
4. "The smaller the anticipated difference in the population, the larger the sample size should be." Is this statement "true" *or* "false"?
5. "Only very large samples can identify very large group differences." Is this statement "true" *or* "false"?
6. Suppose a researcher is planning to conduct a study on attitudes on a controversial topic and expects a wide degree of variation. Given that a wide degree of variation is expected, should the researcher use a "relatively large sample" *or* "a relatively small sample"?
7. "For populations with very limited variability, only very large samples can yield precise results." Is this statement "true" *or* "false"?
8. When studying the incidence of rare phenomena, should researchers use "relatively large samples" *or* "relatively small samples"?
9. Does using a large sample correct for a bias?
10. How do you determine the effect size between two means?

Question for Discussion

11. Think about reports of research in popular media such as TV newscasts. In your opinion, how important is it for the reports to include a mention of the sample size? Explain.

Notes

1. Another way to increase precision is to use stratification. See Chapter 6 for a discussion of stratified random sampling.
2. This is especially true when stratified random sampling is used (as it usually is in national samples) because stratification increases the precision of results over results obtained with simple random sampling.

CHAPTER 8

Standard Error of the Mean and Central Limit Theorem

Chapter Objectives

The reader will be able to:

❑ Describe how sampling error is made.
❑ Explain the concept of the standard error of the mean and the central limit theorem.
❑ Identify how the sample size and variability in a population affect the standard error of the mean.
❑ Describe the 95% or 99% confidence interval and its interpretation.

Suppose there is a large population with a mean of 80 on a standardized test. Furthermore, suppose that a researcher does not have this information but wants to estimate the mean of the population by testing only a sample. When the researcher draws a *random sample* and administers the test to just the sample, will the researcher correctly estimate the population mean as exactly 80? In all likelihood, the answer is "no" because random sampling introduces random (chance) errors – known as **sampling error** as explained in Chapter 7. These errors can affect the results.

At first, the situation may seem rather hopeless. Although bias has been eliminated through the use of random sampling, sampling error

Random (chance) errors are known as **sampling error**.

DOI: 10.4324/9781003299356-10

created by the random selection process may still affect the results. Fortunately, however, random sampling errors have an effect on results that are predictable *in the long run.*

To understand the effects of random sampling error on results in the long run, consider a researcher who drew not just one random sample but many such samples from the population that has a mean of 80. Specifically, the researcher (1) drew a random sample of 60 from a population, tested the participants, and computed the mean for the sample; (2) then drew another random sample of 60 from the same population, tested the participants, and computed the mean for the second sample; (3) then drew a third random sample of 60 from the same population, tested the participants, and computed the mean for the third sample; and (4) continued drawing samples of 60, testing, and computing means an unlimited number of times. In all likelihood, each of these samples would yield somewhat different results because of the effects of random sampling error. Consider these hypothetical results for the first four random samples drawn by the researcher:

In the long run, the effect of random sampling error is predictable.

> *Sample 1*: Mean = 70
> *Sample 2*: Mean = 75
> *Sample 3*: Mean = 85
> *Sample 4*: Mean = 90

First, note that none of the results are correct (i.e., none of the samples have a mean of 80). Second, some of the results are too low (below 80), while some are too high (above 80). Most importantly, note that the average of all four sample means is, in fact, the population mean (i.e., 70 + 75 + 85 + 90 = 320/4 = 80).[1]

The hypothetical researcher would then have a very large number of means (70, 75, 85, 90, and an indefinite number of others), which would create what is known as the *sampling distribution of means.* The **central limit theorem** says that if the sample size is reasonably large, the sampling distribution of means is normal in shape (i.e., forms a normal curve; see Chapter 10 for discussion on shapes of distribution). So if you were to plot the frequencies of these sample means ($\bar{x}$), they would create a normal distribution, where the majority of sample means will be around the actual (true) mean of the population and the rest of sample means will be on the high and low ends in the tails making a normal curve (Figure 8.1).

The **central limit theorem** says that if the sample size is reasonably large, the sampling distribution of means is normal in shape.

Even if the distribution of the sample data behind the means may not be normally distributed, the means of these samples will be. The theorem also states that if the population is normally distributed, the distribution of the sample means of that population will also be normally distributed.

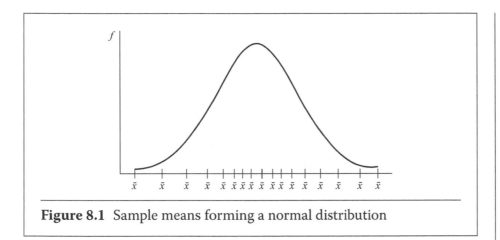

Figure 8.1 Sample means forming a normal distribution

The theorem further states that the mean of all sample means is equal to the actual (true) population mean. The standard deviation of the sampling distribution is known as the ***standard error of the mean*** (SE_M), which as you will see is a very useful statistic. So standard error is essentially the standard deviation of all the means.

For example, let's say a researcher randomly selected a sample of 30 freshmen and calculated the mean of their SAT math scores. Then he repeated this procedure many times. All the sample means from these repetitions are calculated and their frequencies can be plotted, which would create a normal shaped distribution. The mean of these means would be the true mean of the population. And the average of deviations of these means from the overall mean would be the standard error of the means. The theorem is true when the sample size is large (e.g., $n > 30$). And the larger the sample size, the smaller the standard error of the mean will be, generating sample means that cluster closer to the actual population mean. While a discussion of why or how the central limit theorem works is outside the scope of the book, one needs to understand that this theorem is very important since a number of statistics depend on the probabilities generated from the normal distribution of the means.

Of course, in practice, a researcher usually draws a single sample, tests it, and calculates its mean. Therefore, the researcher is not certain of the value of the population mean nor does he or she know the value of the standard error of the mean that would have been obtained if he or she had sampled repeatedly an indefinite number of times. Fortunately, researchers do know two very useful things:

1. The larger the sample, the smaller the standard error of the mean. This is because larger samples have greater precision.[2]

The standard deviation of the *sampling distribution* is known as the ***standard error of the mean*** (SE_M).

2. The less the variability in a population, the smaller the standard error of the mean because sampling from less variable populations yields more precise results.[3]

Given these facts and some statistical theory not covered here, statisticians have developed a formula for estimating the standard error of the mean *(SE$_M$)* based on only the information a researcher has about a single random sample drawn from a population at random.[4] Example 1 shows how the standard error of the mean is sometimes reported after it has been calculated for a set of data.

EXAMPLE 1

"$M = 75$, $s = 16$, $n = 64$,

Formula for Standard Error of Mean: $SE_M = \dfrac{S}{\sqrt{N}}$

Calculation: $\dfrac{16}{\sqrt{64}} = \dfrac{16}{8} = 2$

The standard error of mean is 2. The *SE$_M$ in* this example is what statisticians call the ***margin of error,*** which researchers should use when interpreting the sample mean of 75.[5]

As it turns out, the *standard error of the mean* is a type of standard deviation.[6] As you will see in your study of the standard deviation later in this book, about 68% of the cases in a distribution lie within one standard deviation unit of the mean. Thus, because the standard error of the mean in Example 1 equals 2.00 points, we would expect about 68% of all sample means to lie within 2 points of the true population mean.

Now, consider a practical use of the standard error of the mean. Specifically, it can be used to build a **68% confidence interval for a mean**. To calculate it from the information in Example 1, simply subtract the *SE$_M$* from the mean ($75 - 2 = 73$) and add it to the mean ($75 + 2 = 77$). These two values (73 and 77) are the *limits of the 68% confidence interval for the mean*. In everyday terms, we can say that while the researcher obtained a mean of 75, its value might have been influenced by sampling errors. To account for these errors, a more reliable estimate of the true population mean is that it is between 73 and 77. We can say this with 68% confidence in the correctness of this statement.

TABLE 8.1

Selected Statistics for Two Groups of Participants

	n	Mean	S	95% CI
Group A	128	75.00	16.00	73.59 ~76.41
Group B	200	75.00	10.00	74.29 ~ 75.71

Note: Group A SE_M = 1.41, Group B SE_M = .71

When a researcher reports a single value (such as 75) as an estimate of a population mean based on a sample, it is called a *point estimate* of the mean. When a researcher reports the limits of a confidence interval (such as 73 to 77), it is called an *interval estimate* of the mean. It is usually desirable to report both a point estimate and an interval estimate (or at least report the point estimate as well as the associated standard error of the mean so that a consumer of research can calculate the interval estimate).

It is more common for researchers to report 95% or 99% confidence intervals instead of 68% intervals.[7] When there are several groups, the intervals are often reported in a table such as Table 8.1 in Example 2.

> Confidence intervals of 95% or 99% are commonly reported.

EXAMPLE 2

A 95% confidence interval indicates the range of values within which we have 95% confidence that the true mean lies.

The 95% confidence intervals in Example 2 indicate the range of score values with which we have 95% confidence that the true (i.e., population) mean lies. Consider Group A. We can have 95% confidence that the true mean lies between 73.59 and 76.41.[8] Notice that we are still not absolutely certain of the value of the true population mean because the 128 participants are just a random sample of a population. However, we have a result (the 95% confidence interval) in which we can have a great deal of confidence as an accurate estimate of the population mean.

It should be obvious that small confidence intervals are desirable. As it turns out, the size of the sample is an important factor in the computation of the standard error of the mean. By using reasonably large samples, researchers can minimize the size of the standard error of the mean and thereby obtain confidence intervals that are reasonably small.

It is important to keep in mind that confidence intervals are valid only when the means are obtained with random sampling. If there is bias that

> Using large samples keeps the standard error of the mean and confidence intervals small.

> Confidence intervals are valid only when the means are obtained with random sampling.

creates errors, there are no general techniques for estimating the amount of error created by the bias, nor is it possible to calculate meaningful confidence intervals. This illustrates the importance of using random samples whenever possible.

Exercise for Chapter 8

Factual Questions

1. "If bias has been eliminated, it is safe to assume that the sample is free of sampling errors." Is this statement "true" *or* "false"?

2. Are the effects of random sampling errors predictable in the long run?

3. If a researcher drew several random samples from a given population and measured the same trait for each sample, should he or she expect to obtain identical results each time?

4. "The larger the sample, the larger the standard error of the mean." Is this statement "true" *or* "false"?

5. Suppose a researcher found that $M = 30.00$ and $SE_M = 3.00$. What are the limits of the 68% confidence interval for the mean?

6. Suppose a researcher reported the mean and standard error of the mean. How should you calculate the limits of the 68% confidence interval for the mean?

7. What is the name of the type of estimate being reported when a researcher reports a single value as an estimate of a population mean based on a sample?

8. "It is more common to report the 68% confidence interval than to report the 95% or 99% confidence intervals." Is this statement "true" *or* "false"?

9. How can researchers minimize the size of the standard error of the mean?

Question for Discussion

10. Suppose that you read that the mean of a sample equals 100.00 and the limits of the 95% CI (Confidence Interval) are 95.00 and 105.00. Briefly explain what the limits tell you.

Notes

1. While this example has only four sample means, it illustrates what would happen in the long run. In the long run, with an indefinitely large number of samples, the mean of the sample means obtained by random sampling will equal the true mean of the population. This is not true with biased sampling because a particular bias tends to push all the results off in one direction or the other.
2. See Chapter 7 to review the relationship between sample size and precision.
3. See Chapter 7 to review the relationship between variability and precision.
4. The central limit theorem makes it possible to estimate the standard error of the mean given only one mean and the associated standard deviation.
5. See Appendix A for this and other formulas and computations.
6. In fact, it is an estimate of the standard deviation of the sampling distribution of means.
7. The formulas for building 95% and 99% confidence intervals are beyond the scope of this book.
8. Technically, if a researcher drew 100 samples of 128 participants and constructed 95% confidence intervals for all 100 samples, about 95 of the 100 confidence intervals would include the true mean.

DESCRIPTIVE STATISTICS

As defined in Part A, the goal of descriptive statistics is to summarize data by describing certain characteristics of a given data. Basically, what does the data look like? Descriptive statistics involves organizing a set of numbers by providing a frequency distribution and shapes of distribution for numerical data (normal, skewed), reporting the central tendency (mean, median, or mode), and by how much scores vary from each other (as measured by range, interquartile, or standard deviation). In Part C, details of these descriptive statistics are discussed, as well as z scores in the following sequence.

Chapter 9. Frequencies, Percentages, and Proportions

Chapter 10. Shapes of Distributions

Chapter 11. The Mean: An Average

Chapter 12. Mean, Median, and Mode

Chapter 13. Range and Interquartile Range

Chapter 14. Standard Deviation

Chapter 15. z Score

DOI: 10.4324/9781003299356-11

Frequencies, Percentages, and Proportions

Chapter Objectives

The reader will be able to:

- ❏ Recall the concepts of frequencies, percentages, and proportions.
- ❏ Recognize the calculations of percentages and proportions.
- ❏ Recognize the statistical symbols: f, N.

With any given data, observing the frequencies helps the researcher understand the data better in its initial stages of analysis. *A **frequency** is the number of participants or cases*. Its symbol is f.[1] N, meaning the total *number of participants*, is also used to denote frequency.[2] Thus, if you see in a report similar to Table 9.1, where f = 23 for a score of 99, you would know that 23 participants had a score of 99. Similarly, 20 participants had a score of 100. And N = 43 indicates the total number of participants.

A *percentage*, whose symbol is %, indicates the number per 100 who have a certain characteristic. Thus, if you are told that 44% of the registered voters in a town are registered as Democrats, you know that for each 100 registered voters, 44 are Democrats. To determine how many (*frequency*) are Democrats in a town of 2,200 registered voters (for example), multiply this total number of registered voters by 0.44 (0.44 × 2,200 = 968). Thus, 968 are Democrats.

A **frequency** is the number of participants or cases; its symbol is f. The symbol for *number of participants* is N.

A **percentage** indicates the number per 100.

DOI: 10.4324/9781003299356-12

TABLE 9.1

Distribution of Scores

Score	f
99	23
100	20
	$N = 43$

To calculate a percentage, use division: divide the smaller number by the total number and then multiply by 100. Consider this example: if 27 of 90 gifted children in a sample report are afraid of the dark, determine the percentage by dividing the number who are afraid by the total number of children and then multiply by 100. Thus, 27/90 = 0.30 × 100 = 30%. This result indicates that based on the sample, if you questioned 100 participants from the same population, you would expect about 27 of them to report being afraid of the dark.

A *proportion* is part of one (1) and is usually a decimal number or fraction. In the previous paragraph, the proportion of children afraid of the dark was 0.30, which was the answer obtained before multiplying by 100. This means that *thirty-hundredths* of the children are afraid of the dark. As you can see, proportions are harder to interpret than percentages. Thus, percentages are usually preferred to proportions in all types of reporting. However, do not be surprised if you occasionally encounter proportions in research reports.

When reporting percentages, it is a good idea to also report the underlying frequencies because percentages alone can sometimes be misleading or fail to provide sufficient information. For instance, if you read that 8% of the foreign-language students at a university were majoring in Russian, you would not have enough information to make informed decisions on how to staff the foreign-language department and how many classes in Russian to offer (Table 9.2). If you read that f = 12 (8%), based on a total *of N* = 150 foreign-language students, you would know that 12 students needed to be accommodated.

Percentages are especially helpful when comparing two or more groups of different sizes. Consider the following statistics.

Notice that the total number of foreign-language students (indicated by N) tells us that College B has more Russian majors, but the percentages tell us that College A (with 8%) has more Russian majors *per 100* than College B (with only 4%). Clearly, frequencies and percentages convey different types of information.

To determine a percentage, divide the smaller number by the total and then multiply by 100.

A *proportion* is part of one (1) and is usually a decimal number or fraction.

Report the underlying frequencies when reporting percentages.

TABLE 9.2

Total Number of Foreign-Language Students and Russian Majors

	College A	College B
Total number of foreign-language students	$N = 150$	$N = 350$
Russian majors	$N = 12$ (8%)	$N = 14$ (4%)

Exercise for Chapter 9

Factual Questions

1. What does *frequency* mean?
2. What is the symbol for *frequency*?
3. What does N stand for?
4. If 21% of kindergarten children are afraid of monsters, how many out of each 100 are afraid?
5. Suppose you read that 20% of a population of 1,000 was opposed to a city council resolution. How many are opposed?
6. What statistic is a part of 1?
7. According to this chapter, are "percentages" *or* "proportions" easier to interpret?

Question for Discussion

8. Why is it a good idea to report the underlying frequencies when reporting percentages?

Notes

1. Note that *f* is italicized. If you do not have the ability to type in italics, underline the symbol. This applies to almost all statistical symbols. Also, pay attention to the case. A lowercase *f* stands for *frequency*; an uppercase *F* stands for another statistic, which is beyond the scope of this book.
2. An uppercase *N* should be used when describing a population or total sample size; a lowercase *n* should be used when describing a sample or subgroup.

CHAPTER 10

Shapes of Distributions

Chapter Objectives

The reader will be able to:

❏ Describe normal and skewed distributions of a set of scores.
❏ Recognize the shapes of distributions based on frequency distributions.

The shape of a distribution of a set of scores can be seen by examining a *frequency distribution*, which is a table that shows how many participants have each score. Consider the frequency distribution in Table 10.1. The frequency (i.e., *f*, which is the number of participants) associated with each score (*X*) is shown. Examination of the table indicates that most of the participants are near the middle of the distribution (i.e., near a score of 19) and that the participants are spread out on both sides of the middle with the frequencies tapering off.

The shape of a distribution is even clearer when examining a *frequency polygon*, which is a figure (i.e., a drawing) that shows how many participants have each score. The same data shown in Table 10.1 are shown in the frequency polygon in Figure 10.1. For instance, the frequency distribution shows that three participants had a score of 21; this same

A *frequency distribution* is a table that shows how many participants have each score.

A *frequency polygon* is a drawing that shows how many participants have each score.

DOI: 10.4324/9781003299356-13

TABLE 10.1
Distribution of Depression Scores

Score	*f*
22	1
21	3
20	4
19	8
18	5
17	2
16	0
15	1
	N = 24

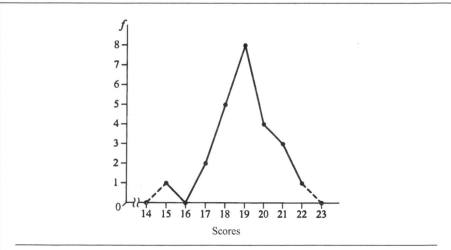

Figure 10.1 Frequency polygon for data in Table 10.1

information is displayed in the frequency polygon. The height of the curve at each score on the horizontal line of the frequency polygon is based on the frequency of each score indicated by the vertical line. The highest point in the polygon shows where most of the participants are clustered (in this case, near the score of 19). The tapering off of the frequency after the score of 19 illustrates how spread out the participants are around the middle.

When there are many participants, thus more numbers, the shape of a polygon becomes smoother and is referred to as a ***curve***. The most important shape is that of the ***normal curve***, which is often called the ***bell-shaped***

A ***curve*** refers to a smoother shape of a polygon.

The ***normal curve*** (also called the **bell-shaped curve**) is the most important curve.

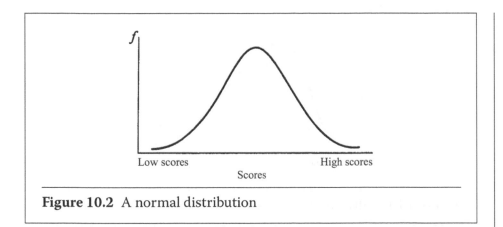

Figure 10.2 A normal distribution

curve. This curve is illustrated in Figure 10.2. The normal curve is important for two reasons. First, it is a shape very often found in nature. For instance, the heights of women in large populations are normally distributed. There are small numbers of extra petite women, which is why the curve is low on the left; many women are of about average height, which is why the curve is high in the middle; and there are small numbers of very tall women on the right side of the curve.

Here is another example of a normal distribution. The average annual rainfall in Los Angeles over the past 110 years has been approximately normal. There have been a very few number of years in which there was extremely little rainfall, many years with about average rainfall, and a very few number of years with a great deal of rainfall.

Another reason the normal curve is important is that it is used as the basis for a number of inferential statistics, which are covered in detail later in Part E.

Some distributions are *skewed.* The curve of the *frequency polygon* is high on the left or the right side. The hump has shifted from the center to either side. For instance, if you plot the distribution of income for a large population, in all likelihood you will find that it has a ***positive skew*** (i.e., is skewed to the right). Examine Figure 10.3. It indicates that there are large numbers of people with relatively low incomes; thus, the curve or the hump is high on the left side. The curve drops off dramatically to the left forming a short tail, with a long tail pointing to the right. This long tail is created by the few numbers of individuals with very high income. Skewed distributions are named for their long tails. On a number line, positive numbers are to the right; hence, the term *positive skew* is used to describe a skewed distribution in which there is a long tail pointing to the right or the positive side (and no long tail pointing to the left). An example of this is inequality of income just discussed. The vast majority of the US

The *normal curve* is used as the basis for a number of inferential statistics.

A distribution with a ***positive skew*** has a long tail to the right.

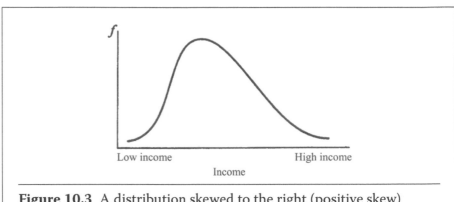

Figure 10.3 A distribution skewed to the right (positive skew)

population is on the lower end of the scale, creating a hump on the lower side of the income scale, and fewer and fewer people on the higher end, creating a long tail to the high (right) end of the income scale.

When the long tail is pointing to the left, a distribution is said to have a *negative skew* (i.e., skewed to the left or the negative side, see Figure 10.4). For example, a negative skew would be found if a large population of individuals was tested on skills in which they have been thoroughly trained. For instance, if a researcher tested a very large population of recent nursing school graduates on very basic nursing skills, a distribution with a negative skew should emerge. There should be large numbers of graduates with high scores, but there should be a long tail pointing to the left, showing that a few number of nurses, for one reason or another—such as being physically ill on the day the test was administered—did not perform well on the test.

Bimodal distributions have two high points. A curve such as that in Figure 10.5 is called bimodal even though the two high points are not

A distribution with a *negative skew* has a long tail to the left.

A *bimodal distribution* has two high points.

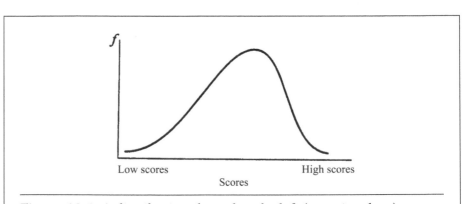

Figure 10.4 A distribution skewed to the left (negative skew)

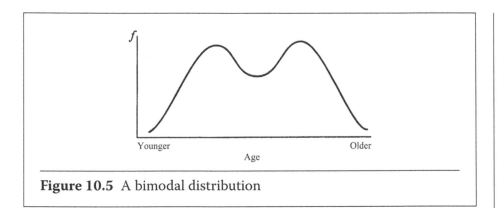

Figure 10.5 A bimodal distribution

exactly equal in height. Such a curve is most likely to emerge when human intervention or a rare event has changed the composition of a population. For instance, if a civil war in a country costs the lives of many young adults, the distribution of age after the war might be bimodal, with a dip in the middle. Bimodal distributions are much less frequent in research than the other types of curves discussed earlier in this chapter.

The shape of a distribution, whether normal or skewed, has important implications for determining which average to compute—a topic that is discussed in Chapters 11 and 12.

> The shape of a distribution has implications for determining which average to compute.

Exercise for Chapter 10

Factual Questions

1. What is the name of a table that shows how many participants have each score?
2. What does a frequency polygon show?
3. What is the most important type of curve?
4. Which type of distribution is often found in nature?
5. In a distribution with a negative skew, is the long tail pointing to the "left" *or* to the "right"?
6. When plotted, income in large populations usually has what type of skew?
7. Suppose that on a 100-item multiple-choice test, almost all students scored between 95 and 100 but a small scattering scored as low as 20. When plotted as a curve, the distribution will show what type of skew?

8. Suppose that a broad cross section of high school students took a very difficult scholarship examination and almost all scored very low, but a very small number scored very high. When plotted as a curve, the distribution will show what type of skew?

9. What is the name of the type of distribution that has two high points?

Question for Discussion

10. Which type of distribution is found most frequently in research than the others?

CHAPTER 11

The Mean

An Average

Chapter Objectives

The reader will be able to:

- ❏ Describe the concept of the mean and its computation and symbols.
- ❏ Use the mean of interval and ratio types of data.
- ❏ Recall the deviations of scores to be the distance of each score from the mean.

When describing the distribution of data, an average is reported as additional information about the data. And the **mean** is the most frequently used average. It is so widely used that it is sometimes simply called the *average*. However, the term *average* is ambiguous because several different types of averages are used in statistics. In this chapter, the *mean* will be considered. Two other types of averages will be considered in Chapter 12.

Computation of the mean is relatively easy: sum (i.e., add up) the scores and divide the total by the number of scores. Here is an example.

Scores: 5, 6, 7, 10, 12, 15
Sum of scores: 55

The **mean** is the most frequently used average.

To compute the *mean*, sum the scores and divide the total by the number of scores.

DOI: 10.4324/9781003299356-14

Number of scores: 6
Computation of mean: 55/6 = 9.166 ≃ 9.17

Notice in the example above that the answer was computed to three decimal places and rounded to two. In research reports, the mean is usually reported to two decimal places.

There are several symbols for the mean. In academic journals, the most commonly used symbols for the mean are *M* and *m*.[1] In addition, some mathematical statisticians use this symbol: $\bar{x}$.

This symbol is pronounced "*x-bar.*" While it is used in some statistics textbooks, it is rarely used in research reports in the social and behavioral sciences.

The ***mean*** is defined as "the balance point in a distribution of scores." Specifically, it is *the point around which all the deviations sum to zero*. An example in Table 11.1 illustrates this characteristic of the mean. The sum of the scores is 60; dividing this by the number of scores (5) yields a mean of 12.00. Subtracting the mean from each score produces the *deviations* from the mean. ***Deviations*** calculate how far each score is away (or deviates) from the mean. Subtracting the mean from each score produces the deviations from the mean. For instance, the first score in Table 11.1 is 7. The score 7 minus the mean 12 yields a deviation of −5. Thus, for the score of 7, the deviation is −5.

The deviations in the last column of Table 11.1 sum to zero.[2] The negative numbers cancel out the positive numbers when summing, yielding zero.

Note that if you substitute any other number in place of the mean and perform the calculations in Table 11.1, deviation scores will not sum to zero. Only the mean of its data will produce this sum. Thus, saying "the mean

Side notes:

M and *m* are the most commonly used symbols for the *mean* in academic journals.

The ***mean*** is the balance point in a distribution of scores.

Deviations calculate how far each score is away (or deviates) from the mean.

The deviations from the *mean* sum to zero.

TABLE 11.1
Scores and Their Deviations from the Mean

Score	Mean*	Deviation**
7	12	−5
11	12	−1
11	12	−1
14	12	2
17	12	5
Sum of scores = 60		Sum of deviations = 0

*Mean = 12 (calculated 60/5), **Deviation is each score minus the mean.

equals 12" is a shorthand way of saying "the value around which the deviations sum to zero is 12."

A major drawback of the mean is that it is pulled in the direction of extreme scores. This is a problem if there are *either* some extremely high scores that pull the mean up *or* some extremely low scores that pull it down. The following is an example of the contributions given to charity by two groups of children, expressed in dollars:

Group A:	1, 1, 2, 3, 3, 4, 4, 4, 5, 5, 5, 5, 6, 6, 6, 7, 8, 10, 10, 10, 11
	Mean for Group A = $5.52
Group B:	1, 2, 2, 3, 3, 3, 4, 4, 5, 5, 5, 6, 6, 6, 6, 6, 9, 10, 10, 150, 200
	Mean for Group B = $21.24

Notice that overall, the two distributions are quite similar. Yet the mean for Group B is much higher than the mean for Group A because two students in Group B gave extremely high contributions of $150 and $200. If only the means for the two groups were reported without reporting all the individual contributions, the mean for Group B would suggest that the average student in that group gave about $21 when, in fact, none of the students made a contribution of about this amount. A distribution that has some extreme scores at one end but not the other is called a *skewed distribution* (see Chapter 10 to review the meaning of a skewed distribution). The mean is almost always inappropriate for describing the average of a highly skewed distribution. An average that provides a more accurate indication of the score of the typical participant in a skewed distribution is described in Chapter 12.

Another limitation of the mean is that it is appropriate only for use with *interval* and *ratio* scales of measurement (see Chapter 3 for a review of scales of measurement). Averages that are appropriate for use with data at the nominal and ordinal levels are described in Chapter 12.

Note that a synonym for *average* is **measure of central tendency**. Although the latter term is seldom used in research reports in academic journals, you may encounter the term in some statistics textbooks.

> The *mean* is pulled in the direction of extreme scores, which can be misleading.

> The *mean* is almost always inappropriate for describing a highly skewed distribution.

> The *mean* should be used only with *interval* and *ratio* scales of measurement.

> A synonym for *average* is **measure of central tendency**.

Exercise for Chapter 11

Factual Questions

1. How is the mean computed?
2. What are the most commonly used symbols for the mean in academic journals?

3. For a given distribution, if you subtract the mean from each score to get deviations and then sum the deviations, what will the sum of the deviations equal?

4. Refer to the example in this chapter of contributions given to charity. Explain why the mean for Group B is much higher than the mean for Group A.

5. If most participants have similar scores but there are a few very high scores, what effect will the very high scores have on the mean?

6. For which scales of measurement is the mean appropriate?

7. The term *measure of central tendency* is synonymous with what other term?

Question for Discussion

8. Is the mean usually appropriate for describing the average of a highly skewed distribution? Why?

Notes

1. The uppercase *M* should be used for the mean of an entire population, and the lowercase *m* should be used for the mean of a sample drawn from a population.

2. Note that if the mean is not a whole number, the sum of the deviations may vary slightly from zero due to rounding when determining the mean because a rounded mean is not *precisely* accurate.

CHAPTER 12

Mean, Median, and Mode

Chapter Objectives

The reader will be able to:

❏ Name the three averages (the measures of central tendency).
❏ State which average is appropriate for different types of measurement scales.
❏ Identify the different values of averages for positive and negative skewed distributions.

You recall that the **mean** described in Chapter 11, is the *balance point* in a distribution. It is the most frequently used average.[1]

An alternative average is the **median**. It is the value in a distribution that has 50% of the cases above it and 50% of the cases below it. Thus, it is defined as the *middle point* in a distribution. In Example 1, there are 11 scores. And 81 is the median for this set of scores. The middle score of 81 has 50% of the numbers on each side, with five scores above 81 and five scores below 81.[2]

The **mean** is the *balance point* in a distribution.

The **median** is the *middle point* in a distribution.

DOI: 10.4324/9781003299356-15

> **EXAMPLE 1**
>
> Scores (arranged in order from low to high):
> 61, 61, 72, 77, 80, 81, 82, 85, 89, 90, 92

In Example 2, there are six scores. Because there is an even number of scores, the median is halfway between the two middle scores. To find the halfway point, sum the two middle scores (7 + 10 = 17) and divide by 2 (17/2 = 8.5). Thus, 8.5 is the value of the median of the set for scores in Example 2.

> **EXAMPLE 2**
>
> Scores (arranged in order from low to high):
> 3, 3, 7, 10, 12, 15

The *median* is not affected by extreme scores.

The advantage of *median* is that it is not affected by extreme scores.[3] This is illustrated in Example 3, in which the extremely high score of 229 has no effect on the value of the median. The median is 8.5, which is the same value as in Example 2, despite the one extremely high score. Thus, the median is insensitive to the skew in a skewed distribution. This illustrates that the median is an appropriate average for describing a highly skewed distribution.

> **EXAMPLE 3**
>
> Scores (arranged in order from low to high):
> 3, 3, 7, 10, 12, 229

The **mode** is the most frequently occurring score or category.

The **mode** is another average. It is defined as the *most frequently occurring score or category*. In Example 4, the mode is 7 because it occurs more often than any other score.

> **EXAMPLE 4**
>
> Scores (arranged in order from low to high):
> 2, 2, 4, 6, 7, 7, 7, 9, 10, 12

EXAMPLE 5

Male and female categories:
M F F M M

In Example 5, the mode is M (males) because it occurs more often than F (females).

There may be more than one mode for a given distribution, which can be a disadvantage when summarizing the data. This is the case in Example 6, in which both 20 and 23 are modes because each occurs twice.

EXAMPLE 6

Scores (arranged in order from low to high):
17, 19, 20, 20, 22, 23, 23, 28

The following are guidelines for choosing among the three averages.

1. Other things being equal, choose the *mean* because more powerful statistical tests described later in this book can be applied to it than to the other averages. However, (1) the mean is *not* appropriate for describing highly skewed distributions, and (2) the mean is appropriate for describing *interval* and *ratio* data, but *not* appropriate for describing *nominal* and *ordinal* data. (See Chapter 3 to review these types of data.)

2. Choose the *median* when the *mean* is inappropriate. Median is appropriate for describing highly skewed distribution. The exception to this guideline is when describing *nominal* data. Nominal data (see Chapter 3) are naming data such as political affiliation, or ethnicity. There is no natural order to these data; therefore, they cannot be put in order, which is required for calculation of the median.

3. Choose the *mode* when an average is needed to describe *nominal* data. Note that when describing nominal data, it is often not necessary to use an average because percentages can be used as an alternative. For instance, if there are more registered Democrats than Republicans in a community, the best way to describe this is to report the percentage of people registered in each party. To

The *mean* is not appropriate for certain types of data.

Choose the *median* when the *mean* is inappropriate, except when describing nominal data.

Choose the *mode* as the average for nominal data. However, for nominal data, reporting an average may not be needed.

TABLE 12.1

A Summary Table of Measures of Central Tendency by Types of Measurement Scale

Type of Scale	Measures of Central Tendency (Mean, Median, and Mode)
Nominal	Mode
Ordinal	Mode/median
Interval/ratio (skewed)	Median
Interval/ratio (normal)	Mean

state only that the modal political affiliation is Democratic (which is the mode in this example) is much less informative than reporting percentages.

Table 12.1 summarizes when mean, median, or mode is appropriate based on the type of the variable. Note that in a perfectly symmetrical distribution, such as the normal distribution, the mean, median, and mode all have the same value (see Figure 12.1).

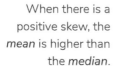
In the normal distribution, the *mean*, *median*, and *mode* have the same value.

When there is a positive skew, the *mean* is higher than the *median*.

When there is a negative skew, the *mean* is lower than the *median*.

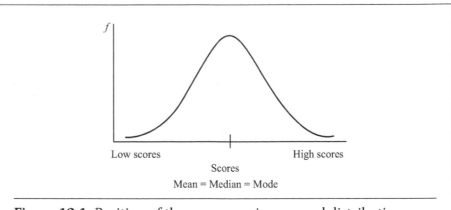

Figure 12.1 Position of three averages in a normal distribution

In skewed distributions, their values are different, as illustrated in Figure 12.2. In a distribution with a positive skew (where the long tail is on the positive side), the mean has the highest value of the three averages because it is pulled in the direction of the extremely high scores. In a distribution with a negative skew (where the long tail is on the negative side), the mean has the lowest value because it is pulled in the direction of the extremely low scores. As noted earlier, the mean should not be used when a distribution is highly skewed.

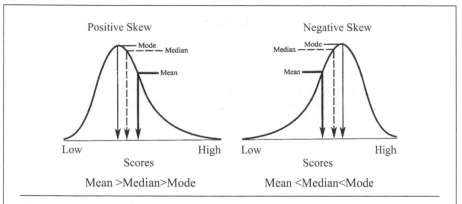

Figure 12.2 Positions of three averages in distributions with positive and negative skews

Exercise for Chapter 12

Factual Questions

1. Which average always has 50% of the cases below it?
2. Which average is defined as the most frequently occurring score?
3. Which average is defined as the middle point in a distribution?
4. If you read that the median equals 42 on a test, what percentage of the participants have scores higher than 42?
5. What is the mode of the following scores:
   11, 13, 16, 16, 18, 21, 25?
6. Is the mean appropriate for describing highly skewed distributions? Why or why not?
7. This is a guideline from this chapter: "Choose the *median* when the *mean* is inappropriate." What is the exception to this guideline?
8. In a distribution with a negative skew, does the "mean" *or* the "median" have a higher value?
9. In a distribution with a positive skew, does the "mean" *or* the "median" have a higher value?

Question for Discussion

10. For describing ordinal data, what is an alternative to reporting the mode where data is expressed in numbers? Give examples of each.

Notes

1. Another term for average is *measure of central tendency*.
2. When there are tie scores in the middle, that is, when the middle score is earned by more than one participant, the method shown here for determining the median is only approximate and the result should be referred to as an *approximate median*.
3. In Chapter 11, it was noted that the mean is pulled in the direction of extreme scores in a skewed distribution, which may make it a misleading average.

CHAPTER 13

Range and Interquartile Range

Chapter Objectives

The reader will be able to:

❏ Recall the statistics that measure the amount of variability in a set of scores called measures of variability.
❏ Explain the range of scores and the limitations when describing data with outliers.
❏ Explain the concepts and computation of the interquartile range, which measures the range of the middle 50% of scores.

In addition to the measures of central tendency (mean, median, and mode), range and interquartile range are used to measure how much scores vary around the mean or median for numerical data.

Variability refers to differences among the scores of participants—how much scores vary from each other.[1] For instance, if all the participants who take a test earn the same score, there is no variability. In practice, of course, some variability (and often quite a large amount of variability) is usually found among participants in research studies.

A group of statistics called *measures of variability* is designed to concisely describe the amount of variability of scores in a set of scores. In this

Variability refers to differences among scores.

Measures of variability are designed to concisely describe the amount of variability in a set of scores.

DOI: 10.4324/9781003299356-16

chapter, two measures of variability (i.e., range and interquartile range) are described. First, however, consider the importance of analyzing variability. The following example illustrates the practical importance of variability:

Suppose a new teacher is going to teach fourth grade next year and is offered a choice between two classes, both of which are very similar in terms of average scores obtained on a standardized test. Before making a choice, the teacher would be wise to ask about the variability. The researcher might learn, for instance, that one class has little variability (the students' scores are all very close to their average), while the other has tremendous variability (the students' scores vary from the highest to the lowest possible with a great deal of spread in between). What does this mean to the teachers making the decision? There is no right or wrong answer to the question, but clearly, information on variability would be important in helping to make a decision and even how the teacher would approach her teaching based on which class she chooses to take on.

A simple statistic that describes variability is the ***range***, which is the difference between the highest score and the lowest score.[2] For the scores in Example 1, the range is 18 (20 minus 2). A researcher could report 18 as the range between the highest and the lowest scores: scores range by 18 points, or simply state that the scores range from 2 to 20.

> The ***range*** is the difference between the highest score and the lowest score.

EXAMPLE 1
Scores: 2, 5, 7, 7, 8, 8, 10, 12, 12, 15, 17, 20

A weakness of *range* is that it is based on only the two most extreme scores, which may not accurately reflect the variability in the entire group. Consider Example 2. As in Example 1, the range in Example 2 is 18. However, there is much less variability among the participants than in Example 1. Notice that in Example 2, except for the one participant with a score of 20, all participants have scores in the narrow range from 2 to 6. Yet, the one participant with a score of 20 has pulled the range up to a value of 18, making it unrepresentative of the variability of the scores of the vast majority of the group.

> A weakness of *range* is that it is based on only the two most extreme scores.

EXAMPLE 2
Scores: 2, 2, 2, 3, 4, 4, 5, 5, 5, 6, 6, 20

Scores such as 20 in Example 2 are known as **outliers**. These lie far outside the range of the vast majority of other scores and increase the size of the range. As a general rule, the range is inappropriate for describing a distribution of scores with outliers.

A better measure of variability is the ***interquartile range (IQR)***. It is defined as the range of the middle 50% of the participants. Using only the middle 50% ensures that the range of the majority of the participants is being described and, at the same time, outliers that could have an undue influence on the ordinary *range* are stripped of their influence.

Example 3 illustrates the meaning of the interquartile range (see Figure 13.1). Notice that the scores are in order from low to high. The arrow on the left separates the lowest 25% from the middle 50%, and the arrow on the right separates the highest 25% from the middle 50%. It turns out that the range for the middle 50% is 3 points. For those interested in the computation of the IQR, notice that the right arrow is at 5.5 and the left arrow is at 2.5. By subtracting $(5.5 - 2.5 = 3.0)$, the approximate IQR is obtained.

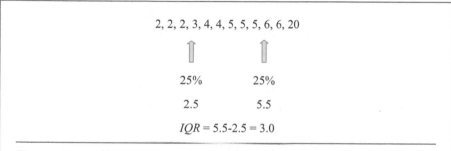

2, 2, 2, 3, 4, 4, 5, 5, 5, 6, 6, 20

| 25% | 25% |
| 2.5 | 5.5 |

$IQR = 5.5 - 2.5 = 3.0$

Figure 13.1 Illustration of the meaning of interquartile range (IQR)

When 3.0 is reported as the *IQR*, consumers of research will know that the range of the middle 50% of participants is 3 points, indicating little variability for the majority of the participants. Note that the undue influence of the outlier score of 20 has been overcome by using the *interquartile range*. In some research, ***semi-interquartile range*** is used, which is half of the interquartile range, measuring the mean of the two middle quartiles.

EXAMPLE 3

Scores: 2, 2, 2, 3, 4, 4, 5, 5, 5, 6, 6, 20

Outliers are scores that lie far outside the range of the vast majority of other scores.

The *interquartile range* (IQR) is the range of the middle 50% of the participants.

IQR is a better measure than *range* because it ignores outliers.

Semi-interquartile range is half of the interquartile range, meaning the mean of the two middle quartiles.

TABLE 13.1
A Summary Table of Measures of Variability

Type of Scale	Measures of Central Tendency (Mean, Median, and Mode)	Measures of Variability (Range, Interquartile, and Standard Deviation)
Nominal	Mode	N/A
Ordinal	Mode/median	N/A
Interval/Ratio (skewed)	Median	Range/interquartile range (semi-interquartile range)
Interval/Ratio (normal)	Mean	Range/standard deviation

When *median* is reported as the average, *IQR* is usually reported as a measure of variability.

The interquartile range may be thought of as the "first cousin" of the *median*.[3] You recall that median is reported for skewed data that may be caused by extreme outliers. (To review median, see Chapter 12.) Thus, when the *median* is reported as the average for a set of scores, it is customary to also report the *interquartile range* as the measure of variability.[4]

As a general rule, it is customary to report the value of an average (such as the value of the median) first, *followed by* the value of a measure of variability (such as the interquartile range).

A summary of measures of central tendency discussed in Chapter 12 and measures of variability (range and interquartile range) are shown in Table 13.1. Standard deviation, which is the most widely used measure of variability, will be discussed in Chapter 14.

Exercise for Chapter 13

Factual Questions

1. What is the name of the group of statistics designed to concisely describe the amount of variability in a set of scores?
2. What are the two synonyms for *variability*?
3. If all participants have the same score on a test, what should be said about the variability in the set of scores?
4. What is the definition of the range?
5. What is a weakness of the range?
6. What is the outlier in the following set of scores: 2, 31, 33, 35, 36, 38, 39?

7. What is the outlier in the following set of scores: 50, 50, 52, 53, 56, 57, 75?

8. As a general rule, is the range appropriate for describing a distribution of scores with outliers?

9. What is the definition of the interquartile range?

10. Is the interquartile range unduly affected by outliers?

11. When the median is reported as the average, it is also customary to report which measure of variability?

Question for Discussion

12. If the differences among a set of scores are great, do we say that there is "much variability" *or* "little variability"?

Notes

1. Synonyms for variability are *spread* and *dispersion*.

2. Some statisticians add the constant one (1) to the difference when computing the range.

3. To calculate the median, count to the middle of the distribution. To calculate the *IQR*, count off the top and bottom quarters. This similarity in computations illustrates why they are "cousins."

4. The measure of variability associated with the mean is introduced in Chapter 14. See Chapter 12 for guidelines on when to report the median and the mean.

CHAPTER 14

Standard Deviation

Chapter Objectives

The reader will be able to:

❑ Describe the concept of standard deviation as how much scores differ from the mean of a group on average.
❑ Identify the 68% range in a normal distribution.
❑ Recall that the higher the standard deviation the higher the variability.
❑ Recognize that standard deviation is reported along with the mean in a normally distributed data.

The **standard deviation** is the most frequently used *measure of variability*. In Chapter 13, you learned that the term **variability** refers to the differences among the scores of participants. Synonyms for *variability* are *spread* and *dispersion*.

The standard deviation is a statistic that provides an overall measurement of how much participants' scores differ from the *mean* score of their group on average. It is a special type of average of the deviations of the scores from their mean.[1]

The **standard deviation** is the most frequently used measure of **variability**.

Variability refers to the differences among the scores of participants. Synonyms for *variability* are *spread* and *dispersion*.

DOI: 10.4324/9781003299356-17

The more spread out the participants' scores are around their mean, the larger the standard deviation. A comparison of Examples 1 and 2 illustrates this principle. Note that S is the symbol for the standard deviation.[2] Note, too, that the mean is the same for both groups (i.e., M = 10.00 for each group), but Group A, with the higher standard deviation among the scores (S = 7.45), has a greater variability than Group B (S = 1.49).

EXAMPLE 1
Scores for Group A: 0, 0, 5, 5, 10, 15, 15, 20, 20 M = 10.00, S = 7.45

EXAMPLE 2
Scores for Group B: 8, 8, 9, 9, 10, 11, 11, 12, 12 M = 10.00, S = 1.49

Now consider the scores in Group C in Example 3. All participants have the same score; therefore, there is no variability. When this is the case, the standard deviation equals zero, which indicates a complete absence of variability of data. Thus, S = 0.00.

EXAMPLE 3
Scores for Group C: 10, 10, 10, 10, 10, 10, 10, 10, 10, 10 M = 10.00, S = 0.00

Considering these three examples, it is clear that the more participants' scores differ from the mean of their group, the larger the standard deviation. Conversely, the less participants' scores differ from the mean of their group, the smaller the standard deviation.

In review, even though the three groups in Examples 1, 2, and 3 have the same mean, the following is true about the variability of the data:

1. Group A has more variability than Groups B and C.
2. Group B has more variability than Group C.
3. Group C has no variability.

Thus, if you were reading a research report on the three groups, you would obtain important information about how the groups differ by considering their standard deviations.

The standard deviation takes on a special meaning when considered in relation to the normal curve (see Chapter 10 to review the normal curve) because the standard deviation was designed expressly to describe this curve. Here is a basic rule to remember. *About two-thirds of the cases (68%) lie within one standard deviation unit of the mean in a normal distribution.* (Note that "within one standard deviation unit" means one unit on *both* sides of the mean.)

Consider this example (illustrated in Figure 14.1): suppose that the mean of a set of normally distributed scores equals 70 and the standard deviation equals 10. About two-thirds of the cases lie within 10 points of the mean. More precisely, 68% (a little more than two-thirds) of the cases lie within 10 points of the mean. As you can see, 34% of the cases lie between a score of 60 and the mean of 70, while another 34% of the cases lie between the mean of 70 and a score of 80. In all, 68% of the cases lie between scores of 60 and 80. This is called the **68% range.**[3]

The 68% rule applies to all normal curves. In fact, this is a property of the normal curve: 68% of the cases lie in the "middle area" bounded by one standard deviation on each side. Suppose, for instance, that for another group, the mean of their normal distribution also equals 70, but the group has less variability with a standard deviation of only 5, which is

> About two-thirds of the cases lie within *one standard deviation unit* of the *mean* in a normal distribution.

> More precisely, 68% (slightly more than two-thirds) of the cases lie within *one standard deviation unit* of the *mean*, also known as **the 68% range.**

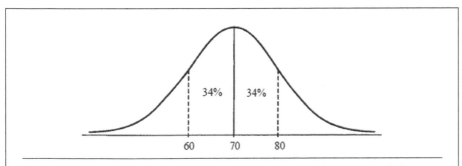

Figure 14.1 Normal curve with a mean of 70 and a standard deviation of 10

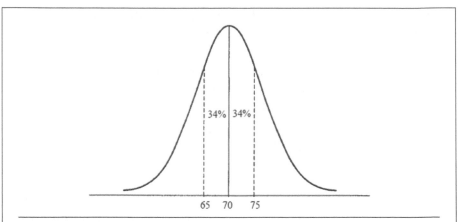

Figure 14.2 Normal curve with a mean of 70 and a standard deviation of 5

illustrated in Figure 14.2. As you can see, 68% of the cases still lie in the middle area of the distribution. However, because the standard deviation is only 5 points, 68% of the cases lie between scores of 65 and 75 for this group of participants.

The 68% guideline (sometimes called *the two-thirds rule of thumb*) strictly applies only to perfectly normal distributions. The less normal a distribution is, the less accurate the guideline is.

By examining Appendix A, you can see that the calculation of the standard deviation is based on the differences between the mean and each of the scores in a distribution. Thus, the standard deviation can be thought of as the first cousin of the mean. As a result, when researchers report the mean (the most frequently used average), they usually also report the standard deviation immediately after reporting the mean. Example 4 shows a sentence in which both the means and standard deviations are reported.

> When the *mean* is reported, the *standard deviation* is also usually reported.

EXAMPLE 4

Sample statement reporting means and standard deviations:

"Group A has a higher mean (M = 67.89, S = 8.77) than Group B (M = 60.23, S = 8.54)."

Exercise for Chapter 14

Factual Questions

1. The term *variability* refers to what?
2. Is the standard deviation a frequently used measure of variability?
3. The standard deviation provides an overall measurement of how much participants' scores differ from what other statistic?
4. If the differences among a set of scores are small, this indicates which of the following (circle one)?
   A. There is much variability.
   B. There is little variability.
5. What is the symbol for the standard deviation when a population has been studied?
6. Will the scores for "Group A" *or* "Group B" below have a larger standard deviation if the two standard deviations are computed? (Do *not* compute the standard deviations; examine the scores to determine the answer.)

Group A:	23, 23, 24, 25, 27, 27, 27
Group B:	10, 19, 20, 21, 25, 30, 40

7. If all the participants in a group have the same score, what is the value of the standard deviation of the scores?
8. What percentage of the cases in a normal curve lies within one standard deviation unit of the mean (i.e., within one unit above *and* one unit below the mean)?
9. Suppose $M = 30$ and $S = 3$ for a normal distribution of scores. What percentage of the cases lies between scores of 27 and 30?
10. Suppose $M = 80$ and $S = 10$ for a normal distribution of scores. About 68% of the cases lie between what two scores?

Question for Discussion

11. If you read the following statistics in a research report, which group should you conclude has the greatest variability? Can you determine the 68% range for these groups?

Group A	$M = 30.23, S = 2.14$
Group B	$M = 25.99, S = 3.01$
Group C	$M = 22.43, S = 4.79$

Notes

1. Appendix A shows how to compute the standard deviation. By studying this appendix, you will learn what is meant by a "special type of average of the deviations."
2. The uppercase, italicized letter S is the symbol for the standard deviation of a population. A Greek letter, sigma (σ), is used as the symbol for the standard deviation of the population. The lowercase, italicized letter s is the symbol for the standard deviation when only a sample of a population has been studied. Note that applied researchers sometimes use $S.D.$ and $s.d.$ as symbols for the standard deviation as well.
3. The percentages, 34% and 68% are approximates, rounded to whole numbers.

CHAPTER 15

z Score

Chapter Objectives

The reader will be able to:

❏ Describe the use of z scores as standardized numbers for comparisons of different data.
❏ Recognize the use of z scores to determine the percentile of a score in a given data.

With a given data, we can create a frequency table and know the distribution of scores, as well as calculate a mean and standard deviation as we have discussed previously. However, oftentimes, data with different scales need to be compared, which may pose a problem. You may have math test scores, for example, measured on a scale of 0–30 and verbal test scores measured on a scale of 0–50. And to compare these two means to see the students' strength area, we couldn't do it using these "raw" scores. They are on different scales not allowing side-by-side comparisons. It would be like comparing apples and oranges because they are on different scales. So, in order to make such a comparison, we need to have some sort of a standard "yardstick" with which to compare them. We can standardize the raw scores in units of standard deviation. These standard scores are called the **z scores**; z scores help us

We can standardize the raw scores in units of standard deviation. These standard scores are called **z scores**.

DOI: 10.4324/9781003299356-18

understand where the scores fall in relation to the other scores in the data. In this way, standardization of scores helps researchers to compare variables with different scales by simply converting raw scores to z scores. A z score is calculated using the data's mean and the standard deviation.

Suppose the mean of the math test scores is 14 with a standard deviation of 4, and you scored 16. This means you scored two points above the mean or half of the standard deviation above the mean.

z scores help us understand where the scores fall in relation to the other scores in the data.

Math test scores: mean = 14, standard deviation = 4
Your math test score = 16

You scored half of the standard deviation above the mean.

Similarly, suppose that the mean of the verbal test scores is 26 with a standard deviation of 10, and you scored 28. This means you scored two points above the mean or one-fifth of the standard deviation above the mean on the verbal test.

Verbal test scores: mean = 26, standard deviation = 10
Your verbal test score = 28

If you compared the raw scores, your verbal test score is higher, but relative to the mean and standard deviation of each test, which of the two test areas would you say, you scored higher?

To convert the raw score to a z score:

Calculate:	$z = \dfrac{\text{raw score} - \text{mean}}{\text{standard deviation}}$
For the math test score:	$z = \dfrac{16 - 14}{4} = \dfrac{2}{4} = 0.5$
For the verbal test score:	$z = \dfrac{28 - 26}{10} = \dfrac{2}{10} = 0.2$

As you can see in the example above, you can take the raw score, subtract the mean, and divide the difference by the standard deviation to produce the z scores. Your z score for the math test is 0.5 and the z score for the verbal test is 0.2. By comparison based on these standardized scores, you scored better on your math test because the z score of 0.5 for the math test score is higher.

In the case that you are working with a population mean instead of a sample mean, you would follow the same steps by taking the difference between the raw score and the population mean and divide by the standard deviation of the population:

$$\text{Calculate: } z = \frac{\text{raw score} - \textit{population} \text{ mean}}{\textit{population} \text{ standard deviation}}$$

In the case that the raw score is lower than the mean, you would end up with a negative *z* score to indicate that the score is below the mean. The mean of all *z* scores in the data would be zero (0) and the standard deviation would always be one (1).

Let us look at the examples of *z* scores below to better understand what these standardized scores mean.

Examples of z scores on a biology test:

Student A	*z* score	=	2.00
Student B	*z* score	=	–1.5
Student C	*z* score	=	1.00

Remember that *z* scores are a yardstick in standard deviation units. Thus, Student A's *z* score of 2.00 would be interpreted as a score that is two standard deviations above the mean; Student B with a *z* score of –1.5 scored one and a half standard deviations below the mean; and Student C's *z* score of 1.00 shows that the score was one whole standard deviation above the mean. In these *z* scores, you learn that Student A and Student C scored above the class mean, and Student B scored below the mean. You can see this by positive and negative *z* scores.

z scores are a yardstick in standard deviation units.

In addition to being able to compare variables of different scales using *z* scores, if the data is normally distributed, one can use the *z* scores to determine the percentile scores of the normal distribution. Consider the standard normal distribution below.

The entire area under the curve totals 100% of the data (see Figure 15.1). Each area in this normal distribution is divided into standard deviation units. The percentages are the total proportion of the normal distribution in each area. Consider placing Student A's *z* score on the normal curve. Student A's standing with the *z* score of 2.00 is in the top 2.27%

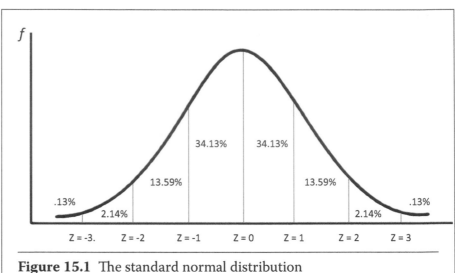

Figure 15.1 The standard normal distribution

(2.14% + 0.13%). In other words, 97.73% (100% − 2.27%) of the students scored below Student A. Thus, we would say, Student A's percentile score or rank is 97.73. **Percentile scores** refer to the percentage of the area that falls below a given score. While measuring the area under all possible z scores is beyond the scope of this book, the important concept to understand is that z scores tell us how high or low a particular score is relative to others in the data through a percentage of scores.

A variety of statistics use z scores. And as discussed earlier, these standardized scores are used to compare scores that are measured in different scales by converting the raw scores to z scores. This way, we can interpret the raw scores in relation to the mean either by understanding the score in terms of standard deviation or by calculating the percentile or the score's rank order.

Percentile scores refer to the percentage of the area that falls below a given score.

The important concept to understand is that z scores tell us how high or low a particular score is relative to others in the data.

Exercise for Chapter 15

Factual Questions

1. What do *z* scores refer to?
2. What do *z* scores help us understand?
3. Suppose a student's *z* score is 3.00. What does this mean?
   A. Discuss in terms of units of standard deviation.
   B. Discuss in terms of its percentile score.

4. How does this student's z score differ from another student whose z score is −3.00?
5. If another student received a test score that showed a z score of 0, what does this mean?
6. The Scholastic Aptitude Test (SAT) total scores range from 400 to 1,600 with a mean of 1,020 and a standard deviation of 194. What is the z score of the score of 1,214? And what does this mean in terms of its percentile score?

Question for Discussion

7. If your raw score produced a negative z score, how is your score below the 50th percentile rank (see Figure 15.1)?

CORRELATIONAL STATISTICS

Correlational statistics is part of descriptive statistics. They describe relationships between two or more variables. While correlational studies do not show causality, they show preliminary information about associations among variables before causality can be determined with further analysis. We will discuss predictability using bivariate and multiple correlations to calculate the coefficient of determination. Correlational statistics discussion in Part D will follow the chapters listed below:

Chapter 16. Correlation
Chapter 17. Pearson *r*
Chapter 18. Scattergram
Chapter 19. Coefficient of Determination
Chapter 20. Multiple Correlation

DOI: 10.4324/9781003299356-19

Correlation

Chapter Objectives

The reader will be able to:

- ❑ Explain that correlation measures how variables are related.
- ❑ Interpret direct (positive) and inverse (negative) relationships.
- ❑ Recognize that correlation does not infer causality.

Still under descriptive statistics, correlation measures associations between variables. ***Correlation*** refers to the extent to which two variables are related across a group of participants. Consider scores on the College Entrance Examination Board's *Scholastic Aptitude Test* (*SAT*) and first-year GPAs in college. Because the *SAT* is used as a predictor in college student selection, there should be a correlation between these scores and first-year GPAs earned in college. Consider Example 1, in which *SAT-V* refers to the verbal portion of the *SAT*.[1] Note that there is one group of students with two numbers for each student *SAT-V* and GPA. Is there a relationship between the two variables?

Upon close observation of numbers in the two variables, you notice that students who scored high on the *SAT-V*, such as Janet and Scotty, had the highest GPAs. Also, those who scored low on the *SAT-V*, such as John

Correlation refers to the extent to which two variables are related across a group of participants.

DOI: 10.4324/9781003299356-20

In a **direct** or **positive relationship**, those who score high on one variable tend to score high on the other, **and** those who score low on one variable tend to score low on the other.

and Hillary had the lowest GPAs. This type of relationship is called a **direct relationship** (also called a *positive **relationship***). In a direct relationship, those who score high on one variable tend to score high on the other, *and* those who score low on one variable tend to score low on the other.

	EXAMPLE 1		

Students	SAT-V	GPA
John	333	1.0
Janet	756	3.8
Thomas	444	1.9
Scotty	629	3.2
Diana	501	2.3
Hillary	245	0.4

Example 2 also shows the scores on two variables for one group of participants. The first variable is self-concept, which was measured with 12 true–false items containing statements such as "I feel good about myself when I am in public." Participants earned one point for each statement that they marked as being true of them. Thus, the Self-concept

	EXAMPLE 2		

Participants	Self-concept	Depression
Sally	12	25
Jose	12	29
Sarah	10	38
Dick	7	50
Matt	8	61
Joan	4	72

scores could range from zero (marking all statements as false) to 12 (marking all statements as true). Obviously, the higher the participant's score, the higher the self-concept is. The second variable is Depression measured with a standardized depression scale with possible scores from 20 to 80. Higher scores indicate more depression.

Do the data in Example 2 indicate that there is a relationship between Self-concept and Depression? Close examination indicates that there is a relationship. Note that participants with high self-concept scores, such as Sally and Jose (both with the highest possible self-concept score of 12), had relatively low Depression scores of 25 and 29 (on a scale from 20 to 80). At the same time, participants with low Self-concept scores, such as Matt and Joan, had high Depression scores. In other words, those with high Self-concepts tended to have low Depression scores, while those with low Self-concepts tended to have high Depression scores. Such a relationship is called an *inverse relationship* (also called a *negative relationship*). In an inverse relationship, those who score high on one variable tend to score low on the other.

In an *inverse* or *negative relationship*, those who score high on one variable tend to score low on the other.

It is important to note that just because a correlation between two variables is observed, it does not necessarily indicate that there is a *causal relationship* between the variables. In Example 2, for instance, the data do not establish whether having a low self-concept causes depression or whether being depressed causes an individual to have a low self-concept. In fact, there might not be any causal relationship at all between the two variables because a host of other variables (such as life circumstances, genetic dispositions, and so on) might account for the relationship between self-concept and depression. For instance, having a disruptive home life might cause some individuals to have low self-concept and at the same time cause these same individuals to become depressed.

Establishing a correlation does not necessarily establish a *causal relationship*.

In order to study *cause-and-effect*, a controlled *experiment* is needed in which different treatments are administered to the participants. (See Chapter 2 for a discussion on experiments.) For instance, to examine a possible causal link between self-concept and depression, a researcher could give an experimental group a treatment designed to improve self-concept and then compare the average level of depression of the experimental group with the average level of a control group.

In order to determine *cause-and-effect*, a controlled *experiment* is needed.

Correlational studies show relationships between variables. Therefore, it is generally inappropriate to infer causality from a correlational study. However, such studies can still be of great value. For instance, going back to the earlier example, if the College Board is interested in how well the *SAT* works in predicting success in college, this can be revealed by examining the correlation between *SAT* scores and college GPAs. It is not necessary for the College Board to examine what causes high GPAs in an experiment to determine the predictive validity of its test.

In addition to validating tests, correlations are of interest in developing theories. Often, a postulate of a theory may indicate that variable X should be related to variable Y. If a correlation is found in a correlational study, the finding helps to support the theory. If it is not found, it calls the theory into question.

Up to this point, only clear-cut examples have been considered. However, in practice, correlational data almost always include individuals who are exceptions to the overall trend, making the degree of correlation less obvious. Consider Example 3, which has the same students as Example 1 but with the addition of two others: Joe and Patricia.

When large numbers of participants are examined, there are almost always exceptions to the trend.

EXAMPLE 3		
Student	**SAT-V**	**GPA**
John	333	1.0
Janet	756	3.8
Thomas	444	1.9
Scotty	629	3.2
Diana	501	2.3
Hillary	245	0.4
Joe	630	0.9
Patricia	404	3.1

To make sense of data with many exceptions, statistical techniques are required.

Joe has a high *SAT-V* score but a very low GPA. Thus, Joe is an *exception* to the rule that high values on one variable are associated with high values on the other. There may be a variety of explanations for this exception: Joe may have had a family crisis during his first year in college *or* he may have abandoned his good work habits to make time for computer games and campus parties as soon as he moved away from home to college. Patricia is another exception; perhaps she made an extra effort to apply herself to college work, which could not be predicted by her *SAT* score. When studying hundreds of participants there will be many exceptions, some large and some small. To make sense of such data, statistical techniques are required. These will be explored in the next chapters.

Exercise for Chapter 16

Factual Questions

1. A direct relationship was found between scores on a reading test and a vocabulary test. This indicates that those who scored high on the reading test tended to have what kind of score on the vocabulary test (circle one)?:
   A. high score.
   B. low score.

2. What is another name for an inverse relationship?

3. Is the relationship between the scores on Test A and Test B "direct" *or* "inverse"?

Participant	Test A	Test B
David	20	600
Julie	30	500
Happy	40	400
Shorty	50	300
Marcia	60	200
Kelly	70	100

4. Is the relationship between the scores on Test C and Test D "direct" *or* "inverse"?

Participant	Test C	Test D
Monica	1	33
Lola	2	38
Jim	4	40
Oscar	6	45
Joey	8	52
Cathleen	9	57

5. There is a positive relationship between the scores on Tests E and F. Which participant is an exception to the rule? Explain why he or she is an exception.

Participant	Test E	Test F
Homer	1050	160
Billy	2508	169
Scott	2702	184
Kathy	3040	205
Leona	5508	90
Bruce	5567	210

6. In an inverse relationship, those who tend to score high on one variable tend to have what kind of score on the other variable?

7. In an inverse relationship, those who tend to score low on one variable tend to have what kind of score on the other variable?

8. What type of study is needed in order to identify *cause-and-effect* relationships?

9. Is *correlation* a good way to determine *cause-and-effect*?

10. When a large number of cases are examined and a positive relationship is found, what else should one expect to find?

Questions for Discussion

11. Name two variables that you think have a direct relationship with each other.

12. Name two variables that you think have an inverse relationship with each other.

Note

1. *SAT-V* scores range from 200 to 800 points.

CHAPTER 17

Pearson *r*

Chapter Objectives

The reader will be able to:

- ❏ Recall that Pearson *r* ranges from −1.00 to 1.00.
- ❏ Identify the direction and strength of a relationship based on the correlation coefficient.

A statistician named Karl Pearson developed a widely used statistic for describing the relationship between two variables (see Chapter 16 for a discussion of the correlation between two variables). The symbol for Pearson's statistic is a lowercase, italicized letter *r*, and it is often called the ***Pearson r***. Its full, formal name is the *Pearson product-moment correlation coefficient*, and there are variations on the name in the research literature, such as the *Pearson correlation coefficient* or the *product-moment correlation coefficient*. These are some of the basic properties of the Pearson *r*:

1. It can range only from −1.00 to 1.00.
2. −1.00 indicates a perfect inverse (negative) relationship, which is the strongest possible inverse relationship.

The full, formal name of the ***Pearson r*** is the *Pearson product moment correlation coefficient*.

DOI: 10.4324/9781003299356-21

A Pearson *r* ranges from −1.00 to 1.00.

Both −1.00 and 1.00 indicate a perfect relationship.

A value of 0.00 indicates the complete absence of a relationship.

3. 1.00 indicates a perfect direct (positive) relationship, which is the strongest possible direct relationship.
4. 0.00 indicates a complete absence of a relationship.
5. The closer a value is to 0.00, the weaker the relationship.
6. The closer a value is to −1.00 or 1.00, the stronger the relationship.

In regard to strength: regardless of the direction (+, −), the higher the *r* is, the stronger the relationship is (see Figure 17.1). For example, an *r* of −0.75 shows a stronger relationship than 0.70. In addition, a higher *r* indicates that there are fewer exceptions to the trend. Inversely, a lower *r* indicates more exceptions to the pattern, and therefore a weaker relationship.

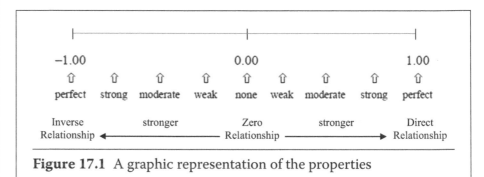

Figure 17.1 A graphic representation of the properties

In regard to direction (+, −): a positive relationship indicates that as the numbers in one variable increase, the numbers in the other variable also increase. For example, with an increase in the number of years of education attainment, there is an increase in salary expectations. In a negative relationship, a higher *r* indicates that as numbers in one variable increase, the numbers in the other variable decrease. For example, as the number of days absent at school increases, test scores decrease.

Let's consider this example. One high school serves two neighborhoods separated by a river. Each neighborhood is vastly different compared to the other in terms of socioeconomic status, with one neighborhood having household income with an average combined income of more than $150,000 (USD) per year and the other neighborhood with an average household combined income of less than $60,000 (USD) per year. A total of 60 high school seniors (30 males and 30 females) representing the two neighborhoods from a class of 300 were selected, and the following results were obtained on their College Entrance Test Scores by their household income.

The result indicates that there is a correlation coefficient of 0.26 between Household Income and College Entrance Test Scores (Table 17.1).

TABLE 17.1

Correlations between Variables

Variables	Household Income	College Entrance Test Scores
Household Income	—	0.26
College Entrance Test Scores		—

This indicates a positive (direct) relationship, which indicates that high Household Income is associated with high College Entrance Test Scores, and low Household Income is associated with low College Entrance Test Scores. So, it is interpreted that with an increase in Household Income there is an increase in College Entrance Test Scores. Said another way, with a decrease in Household Income there is a decrease in College Entrance Test Scores.

Notice that the labels *strong*, *moderate*, and *weak* are used in conjunction with both positive and negative values of the Pearson *r*. Also, note that exact numerical values are not given for these labels. This is because the interpretation and labeling of an *r* may vary from one investigator to another and from one type of investigation to another. For instance, one way to examine **test reliability** is to administer the same test twice to a group of participants without trying to change the participants between administrations of the test. This will result in two scores per examinee, which can be correlated through the use of Pearson *r*. In such a study, a professionally constructed test should yield high values of *r*, such as 0.75 or higher.[1] A result such as 0.65 probably would be characterized as only moderately strong. In another type of study, where high values of *r* are seldom achieved (like predicting college GPAs from College Board scores such as SAT earned a year earlier), an *r* value of 0.65 might be interpreted as strong or even very strong.[2]

In the previous example, while 0.26 on a scale of 0 to 1 doesn't seem very strong, the strength depends on the sample and sample size. Thus, it is difficult to interpret the significance of the correlation by just looking at the correlation coefficient.

Sometimes a Pearson *r* may be misleadingly low if the variability in a group is artificially low or if there is a curvilinear relationship. Appendix B contains some additional notes on the interpretation of values of *r*.

The interpretation of the values of *r* is further complicated by the fact that *r* is *not a proportion*. Thus, for instance, it follows that multiplying 0.50 by 100 does *not* yield a percentage. In other words, 0.50 is *not* equivalent to 50%. This is important because we are used to thinking of 0.50 as

The interpretation of a Pearson *r* varies, depending on the type of study.

Test reliability is examined by administering the same test twice to a group of participants.

An *r* is not a *proportion*. Thus, multiplying it by 100 does *not* yield a percentage.

being halfway between 0.00 and 1.00. Yet, on Karl Pearson's scale for *r*, 0.50, it is *not* halfway between 0.00 and 1.00. This problem in interpretation is explored further in Chapter 32. In that chapter, you will learn about a statistic that is based on *r* and that may be interpreted as a proportion (convertible to a percentage). Appendix A shows the calculation formula for Pearson *r* as a reference.

Exercise for Chapter 17

Factual Questions

1. What is the full, formal name of the Pearson *r*?
2. What does a Pearson *r* of 0.00 indicate?
3. What does a Pearson *r* of –1.00 indicate?
4. Which one of the following indicates the strongest relationship?:
   A. 0.68.
   B. 0.77.
   C. –0.98.
   D. 0.50.
5. Are inverse relationships always weak?
6. Which one of the following indicates the weakest relationship?
   A. 0.93.
   B. –0.88.
   C. –0.95.
   D. 0.21.
7. Is it possible for a relationship to be both direct and weak?
8. Is it possible for a relationship to be both inverse and strong?
9. Consider a value of *r* of 0.65. According to this chapter, would it always be appropriate to characterize the relationship as being "very strong"?
10. Consider a value of *r* of 0.50. Would it be appropriate to multiply this value by 100 and interpret it as representing 50%?

Question for Discussion

11. Very briefly describe a study you might conduct in which it would be appropriate to compute a Pearson *r* (i.e., a study with one group of participants, with two scores per participant). Predict whether the *r* would be "positive" *or* "negative" in direction and whether it would be "high" *or* "low" in strength.

Notes

1. A test is said to be reliable if its results are consistent. For example, if you measured the length of a table twice with a tape measure, you would expect very similar results both times—unless your measurement technique was unreliable.
2. Careful study of the literature on the topic being investigated is needed in order for one to arrive at a non-numerical label or interpretation of a Pearson *r* that will be accepted by other researchers.

CHAPTER 18

Scattergram

Chapter Objectives

The reader will be able to:

- ❏ Analyze Pearson r (a correlation between two variables) illustrated through scattergrams.
- ❏ Recognize the strength (weak to strong and perfect) and direction (+, −) of correlations through patterns in scattergrams.

The term, *correlation*, as discussed thus far, refers to the extent to which two variables are related across a group of participants. (The concept of correlation and Pearson r are described in Chapters 16 and 17, respectively.) The coefficient of determination (r^2), which is used to interpret values of the Pearson r, will be described briefly here and further described in Chapter 19.

As you recall, the Pearson r is a statistic that concisely describes the degree of correlation between two variables, using a single numerical value to describe it. A *scattergram* (also known as a *scatter diagram*) is a statistical figure (i.e., a drawing) that illustrates the correlation between two variables.

A *scattergram* illustrates a correlation between two variables.

DOI: 10.4324/9781003299356-22

MAKING SENSE of Statistics

Coefficient of determination is useful in understanding how much of the unknown in one variable is explained by another variable or variables.

While scattergrams are seldom presented in research reports, by studying the scattergrams in this chapter you will gain a better understanding of the meanings of the values of r (and r^2, which is calculated by squaring the r). The symbol r^2 stands for ***coefficient of determination***, which is useful when interpreting a Pearson r in understanding how much of the unknown in one variable is explained by another variable or variables. To this end, for each scattergram shown in the example here, the associated values of r and r^2 are shown.[1]

Consider the scores in Example 1, which shows scores on a math test taken before the students enrolled in an algebra class and the grades the students subsequently earned in that class. In the example, grades were converted to numbers as follows: 4.0 = A, 3.5 = A-, 3.0 = B, 2.5 = B-, 2 = C, 1.5 = C-, and 1 = D.

EXAMPLE 1		
Student	**Math Test Score**	**Algebra Grade**
Joey	4	1.0
June	6	1.5
Justin	8	2.0
Jill	10	2.5
Janice	12	3.0
Jake	14	3.5
Jude	16	4.0

Notice that the relationship in Example 1 is clearly direct (i.e., positive). Joey has the lowest math score *and* the lowest algebra grade, June has the next-lowest math score *and* the next-lowest algebra grade, and so on. This pattern exists without exception. In fact, the relationship is perfect, so $r = 1.00$ and r^2 also equals 1.00 (i.e., $1.00 \times 1.00 = 1.00$). The values of r^2 can be converted to percentages by multiplying by 100. Doing so, for an r^2 of 1.00 yields 100%. Thus, we can say that 100% of the variance in the algebra grades is accounted for by the variance in the math scores. (See Chapter 19 for a better understanding of the concept of "variance accounted for.") Put in everyday terms, we can say that the data in Example 1 indicate that the basic math scores are a perfect predictor of subsequent algebra grades.

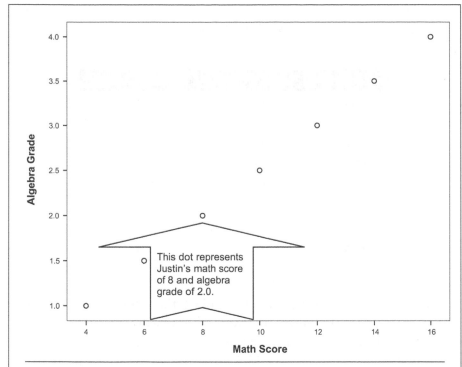

Figure 18.1 Perfect, direct relationship ($r = 1.00$), $r^2 = 1.00$, 100% of variance in Algebra Grade is explained by Math Score.

Figure 18.1 shows the scattergram for the scores in Example 1. Each dot on the scattergram represents the *two* data points for each student. For instance, the dot for Justin shows that he had a score of 8 on the math test and an algebra grade of 2.0 (i.e., a grade of "C") in the algebra class.

Because the relationship is perfect, the dots in Figure 18.1 follow a single straight line. In addition, because the relationship is direct (i.e., positive), the dots go up from the lower-left corner to the upper-right corner.

Perfect relationships are seldom found in the social and behavioral sciences. For instance, in practice, the relationship between math scores and subsequent algebra grades is usually far from perfect because different students have different areas of strength (such as a student being strong in basic math) and weakness (such as the same student being weak in comprehending abstractions in algebra).

Example 2 is the same as Example 1 except that Mike's math score and algebra grade have been added; these are shown in bold in Example 2. Notice that while Mike is near the bottom in his math score, he is about in the middle of the group in algebra grade. In Example 2, Mike is an exception to the overall trend.

When a relationship is perfect, the dots follow a single straight line from the lower left to the upper right.

Perfect relationships are rare.

EXAMPLE 2

Students	Math Score	Algebra Grade
Joey	4	1.0
June	6	1.5
Justin	8	2.0
Jill	10	2.5
Janice	12	3.0
Jake	14	3.5
Jude	16	4.0
Mike	**5**	**3.0**

The relationship is no longer perfect with the addition of Mike's data.

As it turns out, for the data in Example 2, the value of *r* is 0.83, which indicates that the relationship is strong (it applies to all students except Mike), but it is not perfect (because of Mike's exceptional case).

The scattergram in Figure 18.2 portrays the relationship between the scores in Example 2. Note that the overall trend clearly shows that

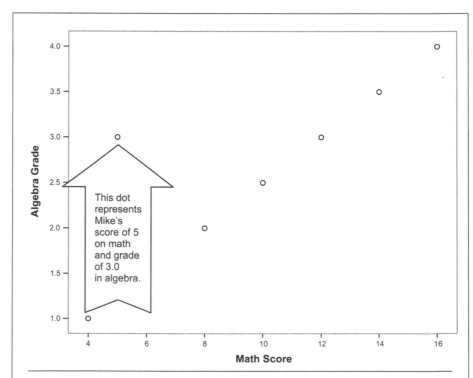

Figure 18.2 Strong, direct relationship (*r* = 0.83), r^2 = 0.69, 69% of variance in Algebra Grade is explained by Math Score.

EXAMPLE 3

Students	Math Score	Algebra Grade
Joey	4	1.0
June	6	1.5
Justin	8	2.0
Jill	10	2.5
Janice	12	3.0
Jake	14	3.5
Jude	16	4.0
Mike	5	3.0
Martin	**13**	**2.0**
Mitch	**11**	**1.5**
Mary	**10**	**2.0**
Michelle	**6**	**1.5**
Manny	**3**	**2.0**

the relationship is direct. However, Mike's pair of scores is an exception, creating a dot that is off the line created by the other dots. Because of Mike's scores, the value of r is now only 0.83 (instead of 1.00 in Example 1, without Mike's case), with a corresponding value for r^2 of 0.69 (for 69% explained variance).

Another example is mother's height and daughter's height. There is a high positive relationship that means tall mothers have tall daughters. However, this is not a perfect relationship because there are exceptions. These exceptions are due to other factors such as the fathers' height and nutrition, just to name a couple of explanations.

When correlating cognitive variables such as math test scores and algebra grades, there will be many exceptions to the trend. Example 3 shows the same scores as in Example 2, with additional students added, each of whom deviates somewhat from the overall trend that indicates a direct relationship. The scores of the additional students are shown in bold in Example 3.

The scattergram in Figure 18.3 shows the relationship between the scores in Example 3. Notice that the overall trend still shows that the relationship is direct (i.e., the dots generally follow a pattern from the lower left to the upper right). However, there is a scattering of dots that indicates deviations from the overall trend. Because of the scatter, r is now only 0.64 (instead of 1.00 in Figure 18.1 and .83 in Figure 18.2). The

When correlating cognitive variables, there will be many exceptions to the trend.

Because of the scatter in Figure 18.3, the relationship is not perfect.

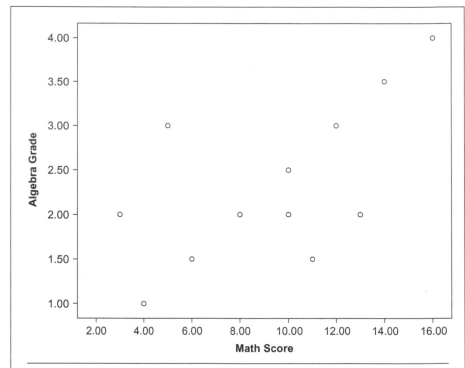

Figure 18.3 Moderately strong, direct relationship ($r = 0.64$), $r^2 = 0.41$, 41% of variance in Algebra Grade is explained by Math Score.

corresponding value of r^2 for an r of 0.64 is 0.41 (for 41% explained variance). Put in everyday terms, the data in Example 3 indicate that the math test is 41% better than zero in the ability to predict algebra grades. Correspondingly, 59% of the variance in algebra grades is unpredicted by the math scores (100% − 41% = 59%).

In comparing the scattergrams thus far, you see that the more scattered the dots, the weaker the relationship. However, the dots in the scattergram that show a pattern going from lower left to upper right (an increase in values in one variable shows an increase in values in the other variable) indicate a direct relationship.

As you know from previous chapters, sometimes relationships are inverse (i.e., negative). Example 4 shows scores on the variables of Cheerfulness and Depression. Cheerfulness was measured with statements such as, "Most mornings when I wake up, I feel cheerful," while depression was measured with statements such as, "I often feel blue on my way to work." Participants responded to the statements on a scale from "strongly agree" to "strongly disagree." Summing their responses across the items, participants could earn scores ranging from zero (marking "strongly disagree" to all items) to 30 (marking "strongly agree" to all items) on each variable.

Examination of the scores in Example 4 suggests an inverse (i.e., negative) relationship. For instance, Sophia and Lauren have the *highest* Cheerfulness scores, but they have the *lowest* Depression scores. At the same time, Luke and Blake have the *lowest* Cheerfulness scores, but they have the *highest* Depression scores. Thus, those who are high on Cheerfulness tend to be low on Depression, while those who are low on cheerfulness tend to be high on Depression.

However, there are exceptions to the overall inverse trend. For instance, while Zack and Bill have the same cheerfulness scores (each has a score of 20), Zack has a much higher depression score (a score of 24) than Bill (a score of 8). These and other exceptions to the inverse trend make the relationship less than perfect, which is illustrated in Figure 18.4 (the scattergram for the data in Example 4).

Example 4 shows scores with an inverse relationship.

EXAMPLE 4

Students	Cheerfulness Score	Depression Score
Sophia	30	4
Lauren	27	5
Joshua	24	7
Zack	20	24
Bill	20	8
Audry	18	10
Richard	16	10
Bobbie	14	14
Kaile	13	9
Kimberly	11	12
Zhi-Chi	9	23
Luke	6	29
Blake	0	30

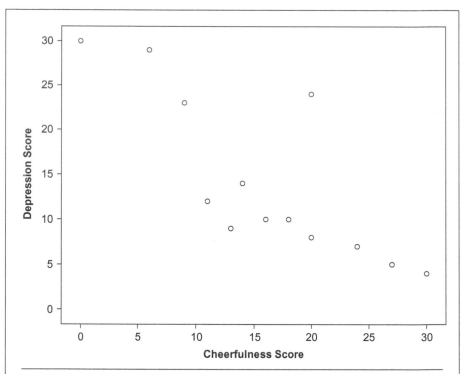

Figure 18.4 Strong, inverse relationship ($r = -0.79$), $r^2 = 0.62$, 62% of variance in Depression Score is explained by Cheerfulness Score.

In Figure 18.4, the pattern from the upper left to the lower right indicates a negative relationship.

The scattering of dots in Figure 18.4 indicates that the inverse relationship is not perfect.

Notice that the overall trend shows that the relationship is inverse (negative) (i.e., the dots generally follow a pattern from the upper left to the lower right). However, there is a scattering of dots (rather than creating a strong line) that indicates that the relationship is not perfect. The value of r for the relationship in Figure 18.4 is −0.79. The corresponding value of r^2 is 0.62 (for 62% explained variance).[2] Put in everyday terms, the data in Example 4 indicate that the Cheerfulness scores are 62% better than zero in the ability to predict Depression scores. Correspondingly, 38% of the variance in depression scores is unpredicted by the Cheerfulness scores (100% − 62% = 38%). Coefficient of Determination will be discussed in Chapter 19 to better explain these percentages.

Concluding Comments

As you can see from the scattergrams in this chapter, the following are true:

- ❏ When the dots form a pattern that goes from the *lower left* to the *upper right*, the relationship is *direct* (i.e., positive).
- ❏ When the dots form a pattern that goes from the *upper left* to the *lower right*, the relationship is *inverse* (i.e., negative).
- ❏ When the dots perfectly follow a single straight line either from the upper left to the lower right or from the lower left to the upper right, the relationship is perfect.
- ❏ The more scattered the dots are around the line, the weaker the relationship.

Exercise for Chapter 18

Factual Questions

1. What is the general term that refers to the extent to which two variables are related across one group of participants?
2. While scattergrams are seldom presented in research reports, they are useful for obtaining what?
3. In Example 1, Jude and Jake have the two highest math scores *and* the two highest algebra grades. This suggests that the relationship is (circle one):
   A. direct.
   B. inverse.
4. "The more scattered the dots are in a scattergram, the stronger the relationship." This statement is (circle one):
   A. true.
   B. false.
5. When a relationship is perfect, what is the percentage of explained variance?
6. "Perfect relationships are frequently found in the social and behavioral sciences." This statement is (circle one):
   A. true.
   B. false.

7. Which of the four scattergrams in this chapter has the greatest amount of scatter?

8. When the dots on a scattergram form a pattern going from the upper left to the lower right, what type of relationship is indicated?

Question for Discussion

9. To what extent do the scattergrams in this chapter and the discussion of them help you to understand the concept of correlation and the meanings of various values of the Pearson *r*? Explain.

Notes

1. Refer to Chapter 31 for a full explanation of coefficient of determination.

2. Note that although the value of *r* is negative, the value of r^2 is positive. When multiplying a negative by a negative (as in squaring a negative), the product is positive. Thus, to determine whether a relationship is negative or positive, the value of *r* must be examined (or the scattergram must be examined). To interpret the strength of *either* a positive or a negative value of *r*, the value of r^2, which is always positive, should be examined.

CHAPTER 19

Coefficient of Determination

Chapter Objectives

The reader will be able to:

- ❏ Recognize the coefficient of determination, which measures the percentage of the variance in one variable accounted for by the variance in the other variable.
- ❏ Recall that computationally the coefficient of determination is r squared.

The **_coefficient of determination_** is useful when interpreting a Pearson r in understanding how much of the unknown in one variable is explained by another variable or variables. Its symbol, r^2 (as described briefly in Chapter 18), explains how it is computed: to obtain it, simply square r. Thus, for a Pearson r of 0.60, r^2 equals 0.36 (0.60 × 0.60 = 0.36).

Although the computation is simple, the meaning of r^2 is sometimes difficult to grasp at first. Let us begin by considering the scores in Example 1. Five young children took an oral vocabulary test before they began learning how to read. After six months of instruction, they took a reading test. As you can see, there is a positive relationship because those who are low on vocabulary (such as John) are also low on reading _and_ those who

To obtain the **_coefficient of determination_**, square r.

DOI: 10.4324/9781003299356-23

are high on vocabulary (such as Diana) are also high on reading. Thus, we can say that the vocabulary scores are predictive of the subsequent reading scores. But how predictive is it?

As it turns out, the value of the Pearson r for the relationship between the two sets of scores in Example 1 is 0.90.[1] Thus, we can say that oral vocabulary scores are highly correlated to reading scores. However, it is possible to be more precise than saying "highly correlated" by using the coefficient of determination in terms of predictability. Let us consider how to be more precise.

The *coefficient of determination*, when converted to a percentage, indicates how much variance on one variable is accounted for by the variance on the other.

EXAMPLE 1		
Students	**Oral Vocabulary**	**Reading**
John	3	6
Janet	5	8
Thomas	4	9
Scotty	9	10
Diana	10	12

First, notice that there are differences among the scores on the oral vocabulary test in Example 1. These differences are referred to as *variance*. There is also variance in the scores on the reading test. When interpreting a Pearson r, an important question is: *What percentage of the variance in one variable is accounted for (or explained) by the other?* If we are trying to predict reading scores from oral vocabulary scores, the question might be phrased as: *What percentage of the variance in reading scores is predicted by the vocabulary scores?* The answer to the question is easily determined. Simply, square r and multiply it by 100. For the scores shown in Example 1, $r = 0.90$. Thus, $0.90 \times 0.90 = 0.81$. And 0.81 in percent is 81%. Thus, the coefficient of determination is 81%.

A percentage may be obtained by multiplying r^2 by 100.

This means that we can determine that 81% of the variance in one variable is accounted for by the variance on the other in this example.[2]

If you draw names at random, your ability to predict is zero percent.

Let us put this into perspective. Suppose we are trying to predict how students will score in reading. Suppose we naively put all the students' names on slips of paper in a hat and draw a name and declare that the first name drawn will probably perform best on the reading test, then draw a second name and declare that this person will probably perform second best on the reading test, and so on. What percentage of the variance in reading will we

If $r = 0.90$, the ability to predict is 81%, better than zero.

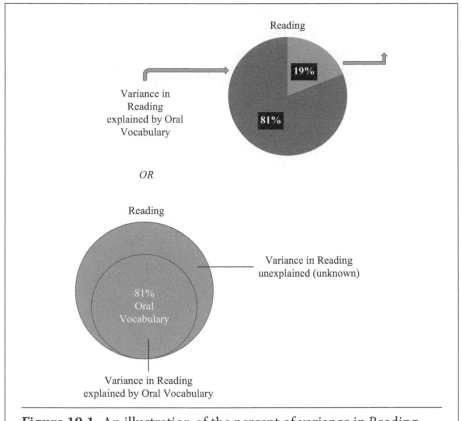

Figure 19.1 An illustration of the percent of variance in Reading
explained by Oral Vocabulary

predict using this procedure? In the long run, with large numbers of students, the answer is about zero (0.00) percent. In the earlier example, the differences (i.e., variance) in oral vocabulary accounted for 81% of the differences (i.e., variance) in reading, which is 81% better than using a random process to make predictions.[3]

As illustrated in Figure 19.1, it follows, however, that if we can account for 81% of the variance, 19% (100% – 81% = 19%) of the variance is *not* accounted for. Thus, we are 19% away from being able to completely predict reading achievement.

Table 19.1 shows selected values of r, the corresponding values of r^2, and the percentage of variance accounted for and not accounted for. Notice that small values of r shrink dramatically when converted to r^2, indicating that we should be very cautious when interpreting small values of r because they are much further from perfection than they might seem at first.[4]

TABLE 19.1

Selected Values of r and r^2

r	r^2	Percent Accounted For	Percent Not Accounted For
0.10	0.01	1%	99%
0.20	0.04	4%	96%
0.30	0.09	9%	91%
0.40	0.16	16%	84%
0.50	0.25	25%	75%
0.60	0.36	36%	64%
0.70	0.49	49%	51%
0.80	0.64	64%	36%
0.90	0.81	81%	19%
1.00	1.00	100%	0%

Small values of r shrink dramatically when squared.

According to Table 19.1, when the values of r are less than 0.40, more than 84% of the variance in one variable is *not* accounted for by the other.

If you are having difficulty understanding the coefficient of determination, consider again what Table 19.1 tells us. When $r = 0.10$, our ability to predict is 1% better than no ability to predict; when $r = 0.20$, our ability to predict is 4% better than no ability to predict, and so on. Thus, the coefficient of determination, when converted to a percentage, tells us how effective one variable is in predicting another expressed in terms of percentages.

Many authors interpret values of r in the 0.20 to 0.40 range as indicating important relationships. While this interpretation may be of some practical importance under certain circumstances, keep in mind that in this r range, 84–96% of the variance on one variable is *not* accounted for by the other. When considering a prediction study, an r of 0.40 (with only 16% of variance accounted for) leaves much room (84%) for improvement when attempting to predict one variable from another.

Exercise for Chapter 19

Factual Questions

1. For a given value of r, how is the value of the coefficient of determination computed?
2. What is the symbol for the coefficient of determination?
3. When $r = 0.50$, what is the value of the coefficient of determination?

4. When $r = 0.50$, what percentage of the variance on one variable is accounted for by the variance on the other?
5. When $r = 0.50$, what percentage of the variance on one variable is *not* accounted for by the variance on the other?
6. When $r = 0.30$, what percentage of the variance on one variable is accounted for by the variance on the other?
7. When $r = 0.30$, what percentage of the variance on one variable is *not* accounted for by the variance on the other?
8. Do "large values" *or* "small values" of r shrink more dramatically when squared?

Question for Discussion

9. When the Pearson $r = 0.40$, is the percentage accounted for equal to 40%? Explain.

Notes

1. Computational procedures for obtaining the value of a Pearson r are beyond the scope of this book, and thus these are shown briefly in Appendix A. In this example, r does not equal 1.00 because there are exceptions to the positive trend; notice that although Janet is higher than Thomas on vocabulary, she is lower on reading.
2. Variance accounted for is sometimes called variance explained.
3. In practice, standardized reading readiness tests designed to predict first-grade reading ability account for only about a third of the variance in reading ability.
4. Note that if there is no variance on either variable, the Pearson r will equal 0.00 and r^2 will also equal 0.00.

CHAPTER 20

Multiple Correlation

Chapter Objectives

The reader will be able to:

❑ Recognize that a multiple correlation coefficient (R) involves the correlation between a combination of variables to increase the predictability of the outcome variable.
❑ Identify that a multiple correlation coefficient has the same basic characteristics as r.
❑ Recall that computationally the coefficient of determination is R squared (multiple correlation coefficient).
❑ Describe that the coefficient of determination (R^2) in percent shows the variance in the outcome variable explained by a combination of variables.

In Chapters 16 through 19, the correlation between only two variables was considered. In this chapter, the correlation between the combination of multiple variables is explored, as well as the *coefficient of determination* of multiple correlations.

To understand the topic of this chapter, consider the three sets of scores in Example 1. (The scores on math scores taken before students

This chapter concerns the correlation between the combination of multiple variables.

DOI: 10.4324/9781003299356-24

enrolled in an algebra class and the algebra grades are the same scores as in Example 3 in Chapter 18, except that they have been rearranged according to algebra grade, with the highest grade of 4.0 at the top.)

EXAMPLE 1

Scores for 13 students arranged according to algebra grades:

Students	Math Score	Attitude Toward Math	Algebra Grade
Jude	16	20	4.0
Jake	14	13	3.5
Janice	12	12	3.0
Mike	5	11	3.0
Jill	10	7	2.5
Martin	13	9	2.0
Manny	3	9	2.0
Mary	10	6	2.0
Justin	8	14	2.0
Mitch	11	8	1.5
Michelle	6	12	1.5
June	6	12	1.5
Joey	4	3	1.0

As you know from Figure 18.3 in Chapter 18, $r = 0.64$ for the relationship between the math scores and subsequent algebra grades, indicating that the math scores are a moderately good predictor of subsequent algebra grades. As you also know, the corresponding value of r^2 is 0.41 (41% explained variance). In everyday terms, this means that 59% of the differences in algebra grades remain *unpredicted* by the math scores.

Because so much of the variance in algebra grades is not predicted by the math scores, it would be desirable to add an additional predictor variable to determine whether the degree of prediction could be improved by using two predictors (instead of using only the basic math scores). To this end, scores on an Attitude Toward Math scale have been added as a potential additional predictor (the middle column).

Attitude Toward Math was measured before enrollment in an algebra class. The Attitude scale had 15 items, with possible scores ranging from 0 to 20 (higher scores indicate a more positive attitude toward math).

Inspection of the Attitude Toward Math scores and the Algebra Grade in Example 1 suggests that there is a direct correlation. For instance, Jude has the highest algebra grade (4.0) and also the highest Attitude Toward Math score (20). In contrast, Joey has the lowest Algebra Grade (1.0) and also the lowest Attitude Toward Math score (3). When high scores on one variable are associated with high scores on the other variable (and when low scores on one are associated with low scores on the other), we know that the relationship is direct (positive).

Inspection of the Attitude Toward Math scores and Algebra Grade Example 1 also suggests that the relationship is not perfect. For instance, Justin has a relatively high Attitude Toward Math score (14) but Algebra Grade of only 2.0 in algebra class. Also, Michelle has a relatively high Attitude Toward Math score (12) but Algebra Grade of only 1.5. These and other exceptions indicate the need to calculate the value of the Pearson r for these data to obtain a precise description of the degree of correlation. As it turns out, the Pearson r for this relationship is 0.68, suggesting that Attitude Toward Math correlates with Algebra Grade.

In review, this is what is known about the ability to predict Algebra Grade based on the data in Example 1:

- ❏ For the relationship between math scores and Algebra Grade, r = 0.64.
- ❏ For the relationship between Attitude Toward Math scores and Algebra Grade, r = 0.68.

Given these relationships, it seems likely that if we use a combination of *both* math scores *and* Attitude Toward Math scores to predict Algebra Grade, the researcher could improve the ability to predict algebra grades over using just one of the predictors.

To determine the degree of relationship between a combination of the two predictors (math scores and Attitude Toward Math scores) to predict Algebra Grade, a ***multiple correlation coefficient***, whose symbol is an uppercase, italicized R, can be calculated. While the calculation of statistics is beyond the scope of this book, it turns out that for the data in Example 1, the multiple correlation coefficient of math scores and Attitude Toward Math on grades is .77 (R = .77) which is higher than only Math Score's correlation to Algebra Grade with an r of 0.64 or only Attitude Toward Math score's correlation to Algebra Grade with an r of .68.

The ability to predict might be improved by using two predictors instead of only one.

Using a combination of two sets of scores might improve the ability to predict.

The symbol for a *multiple correlation coefficient* is an uppercase, italicized *R*.

A ***multiple correlation coefficient*** has the same basic characteristics as *r*.

A multiple correlation coefficient has the same basic characteristics as r. Specifically, the closer the value of R is to zero, the weaker the relationship. For an inverse relationship, the closer R is to -1.00, the stronger the relationship. For a direct relationship, the closer R is to 1.00, the stronger the relationship. Thus, we can say that an R of 0.77 represents a relatively strong, direct relationship because it is fairly close to 1.00.

As you know from Chapter 19, the coefficient of determination for a value of the Pearson r can be calculated by squaring r. Multiplying the square of r by 100% gives the percentage of variance in one variable accounted for by the other variable. The same holds true for R. For the value of R of 0.77 for the data in Example 1, $R^2 = 0.59$ (i.e., $0.77 \times 0.77 = 0.59$). Multiplying 0.59 by 100% indicates that 59% of the variance in algebra grades is accounted for by the combination of Math Score and Attitude Toward Math scores.

Here is a summary of what is now known about the data in Example 1:

❏ The stronger of the two predictors of Algebra Grade is Attitude Toward Math scores ($r = 0.68$, $r^2 = 0.46$ for 46% variance accounted for in algebra grades).

❏ The next predictor of Algebra Grade is Math Score ($r = 0.64$, $r^2 = 0.41$ for 41% variance accounted for in algebra grades).

❏ If *both* Attitude Toward Math scores and Math Score are used *in combination*, the degree of prediction is greater than that for either of the two individual predictors ($R = 0.77$, $R^2 = 0.59$ for 59% variance accounted for in algebra grades).

Clearly, then, for predicting Algebra Grade, it would be better to use a combination of the Attitude Toward Math scores and Math Score.[1]

Note that multiple correlation coefficients can be calculated for a combination of more than two predictors. For instance, it is common to determine the validity of college admissions procedures by calculating the value of R for a combination of at least three predictors, such as (1) verbal college admissions test scores, (2) quantitative college admissions test scores, and (3) high school grades, using freshman grades in college as the variable being predicted. See Figure 20.1 which illustrates a diagram for multiple r.

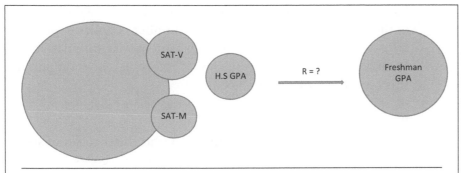

Figure 20.1 Diagrammatic illustration of multiple *r*

Exercise for Chapter 20

Factual Questions

1. Inspection (without any computations) of the Attitude Toward Math scores and the algebra grades in Example 1 suggests that the correlation is (circle one):
   A. direct.
   B. inverse.
2. Inspection (without any computations) of the Attitude Toward Math scores and the algebra grades in Example 1 above suggests that the correlation is (circle one):
   A. perfect.
   B. not perfect.
3. In the example in this chapter, which of the following is the best single predictor of algebra grades (circle one)?
   A. Basic math scores.
   B. Attitude Toward Math scores.
4. Which of the following values of R represents the strongest relationship (circle one)?
   A. $R = 0.45$.
   B. $R = 0.12$.
   C. $R = -0.66$.
5. Suppose a researcher found a value of R of 0.40 for predicting the scores on variable Y from variables X and Z. Expressed as a percentage, what is the amount of variance in variable Y accounted for by the variance in the combination of variables X and Z?

6. Suppose a researcher is examining the validity of a combination of the length of engagement and the number of hours in premarital counseling as predictors of subsequent marital satisfaction. Which correlational statistic should the researcher compute for this research problem (circle one)?

   A. r.

   B. R.

7. Suppose a researcher is examining the validity of a set of scores on an oral language test to predict a set of scores that first-graders will earn on a beginning reading test. Which correlational statistic should the researcher compute for this research problem (circle one)?

   A. r.

   B. R.

8. Can multiple correlation coefficients be calculated for a combination of more than two predictors?

Question for Discussion

9. Examine Example 1. Michelle has the fourth-highest Attitude Toward Math score (a score of 12). However, she has an algebra grade near the bottom of the group (a grade of 1.5). What does this one case tell you about the relationship between Attitude Toward Math and algebra grades? (Keep in mind that the overall relationship is direct.)

Note

1. The value of R is for the best possible combination of the scores on the predictor variables. The statistical methods for determining the best possible combination are beyond the scope of this book.

INFERENTIAL STATISTICS

While descriptive statistics including correlational statistics, discussed thus far, showed ways of describing a sample with its set of numbers, now we are ready to make inferences about a population based on a given sample. This part of the book (Part E) describes inferential statistics, which are statistics that help assess the appropriateness of making inferences from a sample to a population. It is to ask: is what's true about a sample also true about the population?

This involves knowledge about random sampling and the degree to which the sample does or does not represent the population, as discussed in Part B.

Inferential statistics involves hypothesis testing. So, Part E introduces hypothesis testing and how to make decisions about the null hypothesis. Limitations of significance testing, as well as its practical implications are also discussed. The sequence of the topics in Part E is as follows:

Chapter 21. Introduction to Hypothesis Testing
Chapter 22. Decisions about the Null Hypothesis
Chapter 23. Limitations of Significance Testing and Practical Implications

DOI: 10.4324/9781003299356-25

Introduction to Hypothesis Testing

Chapter Objectives

The reader will be able to:

- ❏ Describe the null and alternative hypotheses in their concepts, notations, and interpretations.
- ❏ Apply the concept of directional and non-directional research hypotheses.

In Chapter 1, the word, hypothesis, was introduced and defined as a prediction of the outcomes of research. In other words, a hypothesis is an "educated guess" or an "expected finding." In the context of research, a hypothesis is tested to measure the significance of evidence for a claim about a population using inferential statistics. To begin, typically, two hypotheses are formulated: the *null hypothesis* and the *alternative hypothesis*. The null hypothesis states "no effect," "no difference," or "no relationship." Here are some examples of a null hypothesis.

The **null hypothesis** states "no effect," "no difference," or "no relationship."

The **alternative hypothesis** states that there is a "significant effect," "significant difference," or "significant relationship."

DOI: 10.4324/9781003299356-26

EXAMPLE 1

There is *no difference* in mean math scores between ninth graders and tenth graders after participating in a ten-week math intervention.

EXAMPLE 2

There is *no relationship* between reaction time and age among youth.

EXAMPLE 3

There is *no effect* of early literacy on children's ability to excel in language arts at the elementary age.

Conversely, the alternative hypothesis states that there is a: "significant effect," "significant difference," or "significant relationship." Since the null hypothesis and the alternative hypothesis are direct opposites of each other, if a significant test assumes that the null hypothesis is true, this would make the alternative hypothesis false. And if the null hypothesis is false, this would mean the alternative hypothesis is true.

The null hypothesis assumes no effect, no difference, or no relationship to function as a starting point for the actual outcome. In the absence of no effect, the null hypothesis would be true. Keep in mind that the researcher may or may not agree with the null hypothesis. Nevertheless, it is the starting point, and in most cases, the researcher would want to reject the null hypothesis in the hope of significant findings in their research. So, the researcher sets up the null hypothesis and generally hopes to reject it. This would mean that one can conclude that there is an effect (significant difference or significant relationship) depending on what is being compared. While hypothesis testing is used in a variety of inferential statistics, the remainder of this chapter will focus on hypothesis testing in the context of comparing two means.

Suppose a researcher selected a random sample of first-grade girls and a random sample of first-grade boys from a large school district in order to estimate the average reading achievement of each sample on a standardized test. Further, suppose the researcher obtained these means:

Girls	Boys
$M = 50.00$	$M = 46.00$

This result suggests that girls, on average, have higher achievement in reading than boys. But do they really? Remember that the researcher tested only random samples of boys and girls. Thus, it is possible that the difference the researcher obtained is due only to the errors created by random sampling, which are known as **sampling errors** discussed in earlier chapters. In other words, it is possible that the population mean for boys and the population mean for girls are identical and that the researcher found a difference between the means of the two randomly selected samples only because of the chance errors associated with random sampling. This is the null hypothesis for this research. For the difference between two sample means, it says that: *The true difference between the means (in the population) is zero.*

This statement can also be expressed with symbols as follows:

Null hypothesis:

$$H_0: \mu_1 = \mu_2 \qquad \text{or} \qquad H_0: \mu_1 - \mu_2 = 0$$

where:

H_0 is the symbol for the null hypothesis.

μ_1 is the symbol for the *population* mean for one group (such as all boys).

μ_2 is the symbol for the *population* mean for the other group (such as all girls).

In everyday language, the notation states that the difference between the means of two populations equals zero. Thus, another way to state the null hypothesis is: *There is no true difference between the means.*

There are other ways to express the null hypothesis. For example, the null hypothesis may be stated without the word "no" as follows: *The observed difference between the sample means was created by sampling error.*

Most researchers are interested in identifying differences among groups and are typically seeking explanations for the differences they find. Therefore, most researchers do not undertake their studies in the hope of confirming the null hypothesis. In other words, they usually expect differences (i.e., they hypothesize that differences exist). The researcher's "expectation" is called the **research hypothesis.**

The difference between two means may be due only to *sampling errors*.

For two sample means, *the null hypothesis says that the true difference between the means is zero.*

There are other ways to express the null hypothesis.

A **researcher's expectation** is called the *research hypothesis*.

For instance, a researcher might have a research hypothesis that there is a significant difference between the average reading achievement between girls and boys. So, opposite to the null hypothesis, an alternative hypothesis is set up. That is, the researcher might hypothesize that there is a difference between boys' and girls' reading achievement.

Alternative hypothesis:

$$H_a: \mu_1 \neq \mu_2 \qquad \text{or} \qquad H_0: \mu_1 - \mu_2 \neq 0$$

where:

H_a is the symbol for an *alternative hypothesis* (i.e., an alternative to the null hypothesis)

μ_1 is the symbol for the population mean for one group.

μ_2 is the symbol for the population mean for the other group.

The symbols immediately above indicate that a researcher has a hypothesis (H_a) that states that the mean for one population (μ_1) is not equal ($\neq$) to the mean for the other population (μ_2).

When a study is interested in a significant difference (or no difference), we call this a ***non-directional hypothesis*** (also called a ***two-tailed hypothesis***). The researcher might hypothesize that there is a difference between boys' and girls' reading achievement, but there may be insufficient information to conclude a significant difference.

Suppose the research hypothesis states that one particular group's average is higher than the other group's, this research hypothesis is known as a ***directional research hypothesis*** (also called a ***one-tailed hypothesis***).

A *directional research hypothesis* (H_a) or a *one-tailed test* states that the population mean for Group 1 (μ_1) is higher than the population mean for Group 2 (μ_2). In other words, if the population mean for Group 2 is subtracted from the population mean for Group 1, the difference is greater than zero. The *null hypothesis* (H_0) states that if the population mean for Group 2 is subtracted from the population mean for Group 1, the difference is not greater (the difference is either zero or less than zero).

Expressed as a notation,

Null hypothesis:

$$H_0: \mu_1 \not> \mu_2$$

Alternatively,

Alternative hypothesis:

$$H_a: \mu_1 > \mu_2$$

A ***non-directional research hypothesis*** (also called a ***two-tailed hypothesis***) states that one particular group's mean is different (or not different) from the other group's mean.

A ***directional research hypothesis*** (also called ***one-tailed hypothesis***) states that one particular group's average is higher than the other group's.

where:

H_a is the symbol for an *alternative hypothesis* (i.e., an alternative to the null hypothesis), which in this case is a *directional research hypothesis*.

μ_1 is the symbol for the population mean for the group hypothesized to have a higher mean (in this case, the girls).

μ_2 is the symbol for the population mean for the other group (in this case, the boys).

The symbols directly above indicate that a researcher has a hypothesis (H_a) that states that the population mean for one group 1 is greater than (>) the population mean for the other group (μ_2).

Suppose that the two means we considered at the beginning of this chapter (i.e., $M = 50$ for girls and $M = 46$ for boys) were obtained by a researcher who started with the directional research hypothesis that girls achieve a higher mean in reading than boys. Clearly, the observed means support the research hypothesis, but is the researcher finished? Obviously not, because two possible explanations for the observed difference remain:

1. The observed difference between the samples is solely the result of the effects of random sampling errors. Therefore, there is no true difference. (This is the null hypothesis.) *Or*
2. Girls have higher achievement in reading than boys. (This is the research hypothesis.)

If the researcher stops at this point, there are two explanations for a single difference of four points between the mean of 50 for girls and the mean of 46 for boys. This is hardly a definitive result. The branch of statistics known as *inferential statistics* has statistical techniques (also known as *inferential tests)* that can be used to test the truth of the null hypothesis. If an inferential test allows a researcher to eliminate the null hypothesis, then only the research hypothesis remains, and the researcher can appropriately assert that the data support the research hypothesis.

Consider another example. Suppose a research question asks: Is music at a fast tempo rated as happier than music at a slow tempo by children?

The *null hypothesis* H_0 would state: There is no difference in children's happiness rating between fast tempo and slow tempo music. $H_0 : \mu_1 \not> \mu_2$

The *alternative hypothesis* H_a would state: Music at a fast tempo is rated by children as being happier than music at a slow tempo. $H_a : \mu_1 > \mu_2$

Note that these two hypotheses are mutually exclusive. This means that when one is true, the other is false. Obviously, this means that there is no

The *null hypothesis* is a possible explanation for an observed difference when there are *sampling errors*.

overlap between the null and the alternative hypothesis. This is also true in non-directional hypothesis.

Like it or not, when only random samples have been studied, researchers may be observing differences that are only the result of sampling errors. Thus, the null hypothesis is a possible explanation for any observed difference based on random samples.

Chapter 22 provides information about making decisions about the null hypothesis, and the remaining chapters of this book describe some specific inferential tests of the null hypothesis.

Exercise for Chapter 21

Factual Questions

1. What is the name of the hypothesis which states that a researcher has found a difference between the means of the two randomly selected samples only because of the chance errors associated with random sampling?
2. A researcher's "expectation" is called what?
3. If a researcher believes that Group A will have a higher mean than Group B, is his or her research hypothesis "directional" *or* "non-directional"?
4. Consider this hypothesis, which is expressed in symbols: H_a: $\mu_1 > \mu_2$. Is this a "directional" *or* a "non-directional" hypothesis?
5. What is the symbol for the null hypothesis?
6. What is the symbol for an alternative hypothesis?
7. For what does the symbol μ_1 stand?
8. What is the name of the branch of statistics that has statistical techniques that can be used to test the truth of the null hypothesis?

Question for Discussion

9. A researcher has studied *all* the girls and *all* the boys in the populations of boys and girls in a school. The researcher has found a difference between the mean for boys and the mean for girls. Is the null hypothesis a possible explanation for the difference? Explain.

CHAPTER 22

Decisions about the Null Hypothesis

Chapter Objectives

The reader will be able to:

- ❏ Recognize tests of the null hypothesis or significance tests.
- ❏ Interpret the probability that the null hypothesis is true (*p* value) and its interpretation.
- ❏ Describe type I and type II errors.

As you know from Chapter 21, the null hypothesis states that there is no true difference between two sample means (i.e., in the population, the difference is zero). In other words, it asserts that an *observed difference*[1] between means was obtained only because of sampling errors created by random sampling. The remaining parts (F and G) of this book deal with *tests of the null hypothesis*, which are commonly called *significance tests*.

As its final result, a significance test yields a ***p value***, the probability that the null hypothesis is true. The symbol for probability is an italicized, lowercase *p*. Thus, if a researcher finds, in a given study, that the probability that the null hypothesis is true is less than 5 in 100, this result would be expressed as "$p < 0.05$." How should this be interpreted? What does it indicate about the null hypothesis? Quite simply, it indicates that it is

An inferential test of a null hypothesis yields the *probability* (*p value*) that the null hypothesis is true.

DOI: 10.4324/9781003299356-27

unlikely that the null hypothesis is true. If it is unlikely to be true, what should we conclude about it? We should conclude that it is probably not true.

It is important to understand how to interpret values of *p*, so consider the following analogy. Suppose a weather reporter states that the probability of rain tomorrow is less than 5 in 100. What should we conclude? First, we know that there is some chance of rain— but it is very small. Because of its low probability, most people would conclude that it probably will not rain, and they would not make any special preparations for rain. For all practical purposes, they have rejected the hypothesis that it will rain tomorrow. Likewise, when there is a low probability that the null hypothesis is correct, researchers reject the null hypothesis.

There is always some probability that the null hypothesis is true, so if a researcher waits for certainty, the researcher will never be able to make a decision. Thus, applied researchers have settled on the 0.05 level as the most conventional level at which it is appropriate to make a decision (i.e., decide to reject the null hypothesis, which is done when the probability is 0.05 or less).[2]

When researchers use the 0.05 level, they are, in effect, willing to be wrong 5 times in 100 when rejecting the null hypothesis ($p < 0.05$). Consider the rain analogy earlier. If we make no special preparations for rain during 100 days for which the probability of rain is 0.05, it will probably rain for 5 of those 100 days. Thus, in rejecting the null hypothesis, we are taking a calculated risk that we might be wrong. This type of error is known as a ***type I error*** - the error of rejecting the null hypothesis when it is correct. On those 5 days in 100 when you are caught in the rain without your rain gear, you will get wet because you made 5 type I errors.

A synonym for *"rejecting the null hypothesis"* is "declaring a result statistically significant." In academic journals, statements such as this one are common: *The difference between the means is statistically significant at the 0.05 level.* This statement indicates that the researcher has rejected the null hypothesis because the probability of it being true is 5 or less in 100.

In journals, you will frequently find *p* values of less than 0.05 reported. The two additional common, but more stringent are $p < 0.01$ (less than 1 in 100) and $p < 0.001$ (less than 1 in 1,000). When a result is statistically significant at these levels, investigators can be more confident that they are making the right decision in rejecting the null hypothesis than when they use the 0.05 level. Clearly, if there is only 1 chance in 100 that something is true, it is less likely that it is true than if there are 5 chances in 100 that it is true. For this reason, the 0.01 level is a *higher* level of significance than the 0.05 level, and the 0.001 level is a *higher* level of significance than the

When *p* is less than 5 in 100 that something is true, it is conventional to regard the null hypothesis as *unlikely* to be true.

Applied researchers have settled on the 0.05 level as the most conventional level at which it is appropriate to make a decision, "*p* < 0.05."

A ***type I error*** is the error of rejecting the null hypothesis when it is correct.

A synonym for *rejecting the null hypothesis* is declaring a result *statistically significant*.

TABLE 20.1

Pretest and Posttest Results for Achievement by Site

	Pretest Mean	Posttest Mean	t	Difference
Site A	2.45	3.25	1.732	0.80*
Site B	2.56	2.69	0	0.13
Site C	2.43	4.94	3.65	2.51***
Site D	2.23	4.65	2.60	2.42**

*$p < 0.05$, **$p < 0.01$, ***$p < 0.001$

0.01 level. They are called *higher levels* because at the 0.01 and 0.001 levels, the odds are *greater* that the null hypothesis is correctly being rejected than at the 0.05 level. In statistical reports and journals, these levels are marked by "*" where, commonly, * is used for $p < 0.05$, ** for $p < 0.01$, and *** for $p < 0.001$, and listed in a footnote. It is common to report the probability level in a footnote to indicate statistical significance in inferential statistics. This is illustrated in Table 20.1. It is common to report the probability level in a footnote to an Analysis of Variance (ANOVA) table.

The asterisks in Table 20.1 indicate significances. Accordingly, it reports that Site C shows the most improvement between the pretest and posttest, as the difference is significant at the highest level of 0.001, denoted by three asterisks. Details of these types of comparisons and results will be discussed in later chapters.

The lower the probability (p value), the higher the level of significance. Remember, $p < 0.01$ is a *higher* level of significance than $p < 0.05$.

In review:

0.06+ level: *not* significant, do *not* reject the null hypothesis.
0.05 level: significant, reject the null hypothesis at the .05 level.
0.01 level: more significant, reject the null hypothesis with more confidence than at the 0.05 level.
0.001 level: highly significant, reject the null hypothesis with more confidence than at the 0.01 or 0.05 levels.

So what probability level should be used? Remember that most researchers are looking for significant differences (or relationships). Thus, they are most likely to use the 0.05 level because this is the easiest to achieve of the three levels.[3]

While in inferential statistics, a probability (p) refers to the probability of making a type I error. It is also possible to make a ***type II error***,

It is common to report the probability level in a footnote to indicate statistical significance in inferential statistics.

The lower the probability, the higher the level of significance.

A ***type II error*** is the error of failing to reject the null hypothesis when it is false.

which is the error of failing to reject the null hypothesis when it is false.

EXAMPLE 1

An Example of a Type I Error

The mean self-concept in a population of girls is 50, and the mean self-concept in a population of boys is 50. Thus, the two populations are, on average, equal. A researcher draws a random sample of girls and a random sample of boys and observes a mean of 52 for girls and a mean of 48 for boys. An inferential statistical test indicates that *p is less than 0.05* ($p < 0.05$), so the researcher rejects the null hypothesis. The researcher has unknowingly made a type I error because the researcher has rejected the null hypothesis, which *correctly* states that there is no difference between the population means.

EXAMPLE 2

An Example of a Type II Error

The mean arithmetic reasoning test score in a population of girls is 20, and the mean in a population of boys is 18. Thus, the two populations are, on average, *not* equal. A researcher draws a random sample of girls and a random sample of boys and observes a mean of 21 for girls and a mean of 19 for boys. An inferential statistical test indicates that *p is not less than 0.05* ($p \not< 0.05$), so the researcher does *not* reject the null hypothesis. The researcher has unknowingly made a type II error because the researcher has *not* rejected the null hypothesis, which *incorrectly* states that there is no difference between the population means.

To understand the difference between type I and type II errors, consider Examples 1 and 2.

Carefully considering Examples 1 and 2 will help in understanding the difference between the two types of errors. It is important to note that in both examples, the researcher does *not* know the population means, nor that an error has been made by making decisions based on values of *p*.

In review, there are two types of error that may be made when making a decision about the null hypothesis:

Type I error: reject the null hypothesis when, in reality, it is true.
Type II error: fail to reject the null hypothesis when, in reality, it is false.

The correct decision would be to reject the null hypothesis when, in real-

TABLE 20.2
Type I and II Errors

Decision	When Null Hypothesis is True	When Null Hypothesis is False
Reject null hypothesis	Type I error	
Fail to reject (essentially accept)		Type II error

ity, it is false. Or fail to reject (essentially accept) the null hypothesis when, in reality, it is true. Table 20.2 illustrates this.

At first, this might seem frustrating – researchers have done their best but are still faced with the possibility of making the two types of error. Nevertheless, by using probabilities, researchers make informed decisions in the face of uncertainty. Either decision about the null hypothesis (reject *or* fail to reject) may be wrong, but by using inferential statistics to make the decisions, researchers are making calculated decisions that are consistent with the probabilities.

Because the decision on rejecting the null hypothesis can be wrong in any given case, it is important to examine groups of studies when making generalizations about what is true on any given topic. Across a group of studies on a topic, isolated errors regarding the null hypothesis will be washed out by the majority of other studies. Using probabilities in the way described here makes it very likely that the majority of studies on a given topic are correct in either rejecting or not rejecting the null hypothesis.

Researchers make informed decisions using probabilities in the face of uncertainty.

Exercise for Chapter 22

Factual Questions

1. What does an inferential test of a null hypothesis yield as its final result?
2. Which of the following indicates that the probability is less than 5 in 100 (circle one)?
   A. $p < 0.05$.
   B. $p \nless 0.05$.
3. What is a synonym for the phrase "rejecting the null hypothesis"?
4. What is the name of the error of rejecting the null hypothesis when it is true?
5. If a difference is declared statistically significant, what decision is being made about the null hypothesis?
6. Is the "0.05 level" *or* the "0.01 level" more significant?
7. Is the "0.01 level" *or* the "0.001 level" more significant?
8. When $p < 0.05$, is the difference usually regarded as "statistically significant" *or* "statistically insignificant"?
9. When $p \nless 0.05$, is the difference usually regarded as "statistically significant" *or* "statistically insignificant"?

Question for Discussion

10. Is it possible for a researcher to reject the null hypothesis with absolute certainty?

Notes

1. An "observed difference" is the difference a researcher obtains. It may not represent a true difference because of the influence of random sampling errors.
2. The probability level for rejecting the null hypothesis is known as the *alpha level*.
3. However, if they find that their result is significant at the 0.01 or 0.001 levels, they will usually report it at these levels for the readers' information.

CHAPTER 23

Limitations and Implications of Significance Testing

Chapter Objectives

The reader will be able to:

- ❏ Explain the four basic concepts in significance testing.
- ❏ Demonstrate the steps of a t test for statistical significance.

You will recall the following four important concepts from your study of inferential statistics so far.

1. The *null hypothesis* attributes observed difference between two means to random sampling errors. In effect, it says that any differences observed in random samples (such as the difference between the means of an experimental and a control group) are only chance deviations from a true difference of zero in the population from which the samples were drawn.

2. When there is a low probability that something is true, the researcher rejects it. Thus, if there is a low probability that the null hypothesis is true, such as $p < 0.05$, researchers reject the null hypothesis. (In the topics covered next, in Chapters 24 through 28, you will learn about some specific statistical tests that researchers use to determine the value of the probability for particular types of data.)

DOI: 10.4324/9781003299356-28

3. The lower the p value, the more statistically significant the result, meaning that a researcher can be more confident that he or she is making the correct decision when rejecting a null hypothesis. For instance, if a researcher rejects a null hypothesis at p equal to 0.05, there are 5 chances in 100 that he or she is incorrectly rejecting it. In contrast, if a significance test allows a researcher to reject the null hypothesis at the 0.01 level, he or she is taking only 1 chance in 100 that the decision to reject it is incorrect.

4. Lastly, when a null hypothesis has been rejected the difference is *statistically significant*.

Knowing these four basic concepts in significance testing will help you to understand this chapter of this book.

Typically, researchers hope to find that their differences are statistically significant (i.e., *reliable*, because it is unlikely that they are due only to random sampling). Furthermore, they hope that they will be highly significant at levels such as 0.01 or 0.001. This allows them to state in their research report that the differences they have observed in their research are highly reliable. Reliable results are ones that the researcher can count on from observation to observation or study to study. Consider this analogy: if an employer says that an employee is "reliable," the employer means that the employee can be counted on, time after time, to perform their work in a consistent manner. Just as employers are seeking reliable employees, researchers are seeking to identify reliable differences.

Unfortunately, some researchers make the mistake of equating a significant (i.e., reliable) difference with a "large difference." However, just because a difference is reliable does not necessarily mean that it is a large difference. Consider Example 1, which illustrates that a small difference can also be a reliable difference.

EXAMPLE 1

An individual notices that the number of minutes of daylight is slightly greater on December 22 than on December 21 (the shortest day of the year). She decides to sample the last 75 years and make the same measurements. Year after year, she obtains the same small difference when comparing the number of minutes of daylight on December 21 with the number on December 22. Conducting a t test on the difference between the average number of minutes on December 21 and December 22, she finds that she has identified a *statistically significant* (i.e., reliable) difference. Note, however, that while the difference is reliable, it is quite small.

*Researchers typically hope to identify **statistically significant** differences.*

Just because a difference is statistically significant does not mean that it is a large difference.

There are three factors that are mathematically combined to determine the significance of the difference between two means. They are (1) the size of the mean difference, (2) the size of the sample, and (3) the amount of variation from one observation to another. While the size of the difference in daylight in Example 1 is small, the size of the sample is reasonably large ($n = 75$). More important, there is essentially no variation from year to year (i.e., the number of minutes of daylight on December 21 is the same for each of the 75 years). This lack of variation indicates that a highly reliable (i.e., statistically significant) difference has been observed (even though it is a small difference).

Example 1 is an extreme example designed to help you grasp the concept that *even a small difference can be statistically significant*. It is important to note that small statistically significant differences are frequently reported in all the sciences, especially in the social and behavioral sciences. Thus, for consumers of research, it is not sufficient to know that a difference is statistically significant; they also need to know the size of the difference. It is not uncommon for a researcher (or someone who is writing about another's research) to state that a variable mean is statistically significantly higher than another variable mean without discussing the magnitude of the difference—as though being statistically significant is equivalent to being large.

> Even a small difference can be statistically significant.

In practical terms, it is usually true that a small difference (even though it is significant) is less likely to be of practical importance than a large difference. However, it is also true that *even a small, significant difference can sometimes be important*.

> Even a small, significant difference can sometimes be important.

For instance, suppose that a researcher found that making only a very modest change in diet created a very small decrease in the side effects experienced by individuals undergoing chemotherapy. Suppose that the difference was statistically significant, indicating that it is a reliable difference. Is this small difference of practical importance? Arguably, those who are experiencing the side effects would probably find any degree of relief from side effects to be very important, especially because it involves only a very modest (and presumably inexpensive) change in diet.

In light of this, you may be wondering why researchers conduct significance tests. The answer is clear if you keep in mind that the evaluation of a difference is a *three-step process* in which significance testing comes first. To make the discussion easy to follow, consider an experiment in which there is an experimental group and a control group, both of which were drawn at random. Assume that the difference between the mean of the experimental group and the mean of the control group is being tested with a *t* test for statistical significance. (This test will be explained at length in the next chapters.) In order, the steps are:

> Evaluating a difference is a *three-step process* in which significance testing comes first.

Step 1. Use a significance test to determine if the difference is statistically significant (i.e., reliable and unlikely to be due to chance). If it is *not* statistically significant, proceed to Step 2. If it is statistically significant, skip Step 2 and proceed to Step 3.

Step 2. For statistically *not* significant results, the researcher should *not* assert that the experimental treatment is superior to the control condition. At this point, a researcher might abandon the hypothesis underlying the research or decide to conduct a more rigorous experiment with a larger number of participants. Note that if the treatment for the experimental group is one that is already in use (say, an unproven herbal treatment for depression that is already being advertised and sold in specialty stores), the result might have ***practical implications***. In this instance, the researcher might want to assert that, within the limitations of their study, a reliable effect was *not* found and that individuals probably do *not* benefit from consuming the herbal treatment. The practical implication would be that consumers probably should stop purchasing the herbal treatment. Thus, even an insignificant difference might have a practical implication.

Step 3. For a statistically significant result, evaluate the statistically significant difference in terms of its *practical implications or practical significance*. In order to consider its practical significance, a researcher will need to consider the size of the difference[1] in relation to the benefit. In an experiment, the ideal is to find an inexpensive treatment (such as taking a small dosage of aspirin regularly) that produces a large beneficial effect (say, a very large reduction in the incidence of heart attacks). However, if the treatment produces only a small (but statistically significant) difference, the cost of using the treatment becomes an important consideration. For instance, if an experiment revealed that a slight modification in the way teachers discuss the concept of simple fractions (an inexpensive treatment) produces, on average, a small increase in students' scores on a test on simple fractions, the treatment might be of practical value. (In other words, a very inexpensive treatment that produces a small, beneficial difference might be worth pursuing.) In contrast, if a small difference (such as a small difference in achievement test scores) comes at a very high price (such as having to purchase expensive computer-assisted instructional software for each student), the cost might mitigate the practical significance of the results.

A statistically *not* significant difference can have ***practical implications***.

Consider the size of the difference when evaluating a statistically significant difference.

Consider the cost in relation to the benefit when evaluating a statistically significant difference.

Small significant differences might be worth pursuing if the cost is low. If the cost is high, they might be of limited practical significance.

Considering the second and third steps discussed here determining practical significance is a complex process that should be undertaken after a significance test is conducted.

In conclusion, there are two fundamental limitations of significance testing. Failure to recognize them can lead to misinterpretation of research results: (1) it fails to indicate the size of a difference, and (2) it does not assess the practical significance of a difference. Unfortunately, some researchers and reviewers who summarize and report on the research of others fail to recognize these limitations and assume that all statistically significant differences are inherently large and therefore of practical importance.[2] Their failure can lead them to misinterpret the results of research and make poor decisions that they claim are "research-based."

In Chapter 22 and this chapter, the emphasis has been on using probabilities in comparing means because this is one of the most common uses for inferential statistics. Chapters 24 to 27 provide more information on statistical tests of significance for means.[3]

There are two fundamental limitations of significance testing. Failure to recognize them can lead to misinterpretation of research results.

Exercise for Chapter 23

Factual Questions

1. To what does the null hypothesis attribute differences?
2. When should the null hypothesis be rejected (circle one)?
   A. When the probability (p) is low.
   B. When the probability (p) is high.
3. Is the statement that "the difference is statistically significant" completely equivalent to saying "the difference is large"?
4. Can a small difference be statistically significant?
5. Can a small significant difference sometimes be important?
6. Is it possible for a "no significant" difference to have practical implications/significance?
7. In an experiment, what is the "ideal" finding regarding cost in relation to benefit?
8. Should practical significance be determined before statistical significance is determined?
9. According to this chapter, is determining practical significance a complex process?

Question for Discussion

10. Suppose a student made this statement: "Research shows that treatment with Alpha is significantly better than the Beta treatment." How would you respond to this student? Would you ask for additional information based on what you learned from this chapter? Explain.

Notes

1. See Chapter 32 for a discussion of how statisticians standardize the quantification of the size of the differences between pairs of means.

2. As you know from Chapter 21, it is true that the larger the difference between two means, the more likely that it is statistically significant. However, as you also know, the sample size and the amount of variation contribute to the decision regarding statistical significance. Failure to recognize that *three* factors contribute to the decision regarding statistical significance (not just the size of the difference) probably contributes to some of the confusion and misinterpretation of the results of significance testing.

3. All other descriptive statistics may be compared for significance through use of inferential statistical tests. For instance, there are tests for comparing medians, standard deviations, and correlation coefficients. Although the mathematical procedures are different, in any such comparison there is a null hypothesis that attributes differences to sampling errors. Decisions about the null hypothesis for these and other descriptive statistics are made using the same probability levels discussed in this chapter.

MEANS COMPARISON

Having introduced hypothesis testing, null hypothesis, and how we make decisions about the null hypothesis in Part E, the remaining two parts, F and G, take a closer look at hypothesis testing through various types of means comparisons and predictive analysis. In particular, Part F introduces the t test, followed by three types of t tests to compare two means: Independent Samples t Test, Dependent Samples t Test, and One-Sample t Test, as well as various ways of reporting t test results including sample results. One-Way and Two-Way ANOVA are also discussed. The sequence of the topics in Part F is as follows:

Chapter 24. Introduction to the t Test
Chapter 25. Independent Samples t Test
Chapter 26. Dependent Samples t Test
Chapter 27. One-Sample t Test
Chapter 28. Reporting the Results of t Tests: Display of Outcomes
Chapter 29. One-Way ANOVA
Chapter 30. Two-Way ANOVA

DOI: 10.4324/9781003299356-29

CHAPTER 24

Introduction to the *t* Test

Chapter Objectives

The reader will be able to:

- ❏ Explain the concept of the *t* test as a test of the difference between two sample means to determine statistical difference.
- ❏ Identify the factors affecting the probability level that the null hypothesis is true.
- ❏ Differentiate the difference among independent samples, dependent samples, and One-Sample *t* Tests.

Researchers frequently need to determine the statistical significance of the difference between two sample means. Consider Example 1 in this chapter, which illustrates the need for making such a comparison.

DOI: 10.4324/9781003299356-30

> **EXAMPLE 1**
>
> **An example of Independent Samples *t* Test**
>
> A researcher wanted to determine whether there are differences between men and women voters in their attitudes toward welfare. Separate samples of men and women were drawn at random and administered an attitude scale. Women had a mean of 38 (on a scale from 0 to 50, where 50 was the most favorable attitude). Men had a mean of 35. The researcher wanted to determine whether there is a significant difference between the two means. What accounts for the three-point difference between a mean of 38 and a mean of 35? One possible explanation is that the population of women has a more favorable attitude than the population of men and that the two samples correctly reflect this difference between the two populations. Another possible explanation is raised by the *null hypothesis*, which states that there is no true difference between men and women—that the observed difference is due to sampling errors created by random sampling.

The *t test* tests the difference between two sample means to determine significant difference.

When the *t test* yields a low probability that a null hypothesis is correct, researchers usually reject the null hypothesis.

Basic Factor 1. The larger the samples, the less likely that the difference between two means was created by sampling errors.

While Example 1 illustrates an Independent Samples *t* Test, which will be explained in great detail in the next chapter, this chapter deals with how to compare two sample means for statistical significance in general.

About 100 years ago, a statistician named William Gosset developed the *t test* for exactly the situation described in Example 1 (i.e., to test the difference between two sample means to determine whether there is a significant difference or statistical significance between them).[1] As a test of the null hypothesis, the *t* test yields a probability that a given null hypothesis is correct. As indicated in Chapter 22, when there is a low probability that it is correct—say, as low as 0.05 (5%) or less—researchers usually reject the null hypothesis.

The computational procedures for conducting *t* tests are beyond the scope of this book.[2] However, the following material describes what makes the *t* test work. In other words, what leads the *t* test to yield a low probability that the null hypothesis is correct for a given pair of means? Here are the three basic factors that interact with each other in determining the probability level:

Basic factor 1. The larger the samples, the less likely that the difference between two means was created by sampling errors. This is because larger samples have less sampling error than smaller samples. Other things being equal, when large samples are used, the *t* test is more likely to yield a probability low enough to allow rejection of the null hypothesis than when small samples are used. Put another way: when there is less sampling error, there is less chance that the null hypothesis (an assertion that sampling error created the difference) is correct. In other words, when there is great precision because large samples are used, researchers can be more confident that their sample results reflect the underlying difference in a population than when small samples are used.

Basic factor 2. The larger the observed difference between the two means, the less likely that the difference was created by sampling errors. Random sampling tends to create many small differences and few large ones (especially if large samples are used). Thus, when large differences between means are obtained, the *t* test is more likely to yield a probability low enough to allow rejection of the null hypothesis than when small differences are obtained. Said another way, the larger the observed difference, the more likely the researcher will be able to reject the null hypothesis.

Basic factor 3. The smaller the variance among the participants, the less likely it is that the difference between two means was created by sampling errors and the more likely the null hypothesis will be rejected. To understand this principle, suppose a population of people has no variance because everyone is identical: all members of the population look alike, think alike, and speak and act in unison. How many participants does the researcher have to sample from this population to get an accurate result? Obviously, only one participant because all members of the population are identical. Thus, when there is no variation among the members of a population, it is not possible to have sampling errors when sampling from the population. If there are no sampling errors, the null hypothesis should be rejected. As the variation increases, sampling errors are more and more likely to be the cause of an observed difference between means.[3]

There are three most common types of *t* tests: an ***Independent Samples t Test*** for *independent data* (sometimes called *uncorrelated data*), a ***Dependent or Paired Samples t Test*** for *dependent data* (sometimes called *correlated data*), and a ***One-Sample t Test*** for one sample that is compared to a population mean. As stated earlier, Example 1 is an example of an Independent Samples *t* Test. It has independent data, which means that there is no pairing or matching of individuals across the two samples.

Men and women were drawn independently from the two populations without regard to whether each individual in one group "matches" in any way (such as age) each individual in the other group.

Basic factor 2. The larger the observed difference between the two means, the less likely that the difference was created by sampling errors.

Basic factor 3. The smaller the variance, the more likely the null hypothesis will be rejected.

When there is no variation, it is not possible to have sampling errors when sampling from the population.

Independent Samples t Test is used for independent data.

Dependent (paired) Samples t Test is used for dependent data.

One-Sample t Test is used for one sample.

The meaning of independent data becomes clearer when contrasted with dependent data in a dependent (or paired) samples t test, which is illustrated in Example 2.

EXAMPLE 2

An example of Dependent Samples t Test

In a study of visual acuity, pairs of same-sex siblings (two brothers or two sisters) were identified for a study. For each pair of siblings, a coin was tossed to determine which one received a vitamin supplement and which one received a placebo. Thus, for each participant in the experimental group, there was a same-sex sibling in the control group.

Example 2 illustrates a Dependent or Paired Samples t Test.

The means that result from the study in Example 2 are less subject to sampling error than the means from Example 1. Remember that in Example 1 there was no matching or pairing of participants before the assignment to conditions. In Example 2, the matching of participants assures us that the two groups are more similar than if just any two independent samples were used. To the extent that genetics and gender are associated with visual acuity, the two groups in Example 2 will be more similar at the onset of the experiment than the two groups in Example 1. The t test for dependent data takes this possible reduction of error into account.[4] Researchers will often mention in their reports whether they conducted t tests for independent or for dependent data.

A One-Sample t Test compares a sample mean to a population mean or a historical mean. Consider this example for a One-Sample t Test.

Dependent data may have less sampling error.

EXAMPLE 3

An example of One-Sample t Test

In a study comparing the national mean score of students on the Mathematics Assessment Test of the National Assessment of Educational Progress (NAEP) to a group of students at a particular school, 250 students are randomly selected. The calculated mean from this sample is compared with the mean of the population to see if there is a difference.

TABLE 24.1
A Summary Table of Means Comparison *t* Tests

Means Comparison *t* Test	Samples	Comparison	Independent (Grouping) Variable	Dependent (Test) Variable
Independent Samples *t* Test	Two independent samples	Two independent sample means	Categorical data (with two categories)	Continuous data (normally distributed)
Dependent Samples *t* Test	One sample with repeated measures *or* two related samples	Two repeated measures means *or* two dependent sample means	Categorical data (with two categories)	Continuous data (normally distributed)
One-Sample *t* Test	One sample with one measure	A sample mean and a population mean	N/A	Continuous data (normally distributed)

In this example, the sample mean from a particular school is compared with the national mean (the population). Table 24.1 summarizes various *t* tests for means comparison. Information in the table will be explained in detail in the next chapters.

Independent (grouping) variable and dependent (test) variable will be discussed in the next chapters as these three types of mean comparison *t* tests are discussed in depth, with further elaboration on the examples introduced in this chapter.[5]

A new statistical notation needs to be introduced in this context of means comparison. The notation, *df*, stands for *degrees of freedom*. The degrees of freedom count the number of independent pieces of information used for the result (N) minus the number of groups (for example, $N - 1$ degrees of freedom for one sample group). In the case that *df* = 10, this shows that there was an *N* of 11 in the study.

Degrees of freedom count the number of independent pieces of information used for the result (*N*) minus the number of groups.

Exercise for Chapter 24

Factual Questions

1. Example 1 mentions how many possible explanations for the three-point difference?
2. What is the name of the hypothesis that states that the observed difference is due to sampling errors created by random sampling?

3. Which of the following statements is true (circle one)?
   A. The t test is used to test the difference between two sample means to determine statistical significance.
   B. The t test is used to test the difference between two population means to determine statistical significance.
4. If a t test yields a low probability, such as $p < 0.05$, what decision is usually made about the null hypothesis?
5. The larger the sample, the (circle one):
   A. more likely the null hypothesis will be rejected.
   B. less likely the null hypothesis will be rejected.
6. The smaller the observed difference between two means, the (circle one):
   A. more likely the null hypothesis will be rejected.
   B. less likely the null hypothesis will be rejected.
7. If there is no variation among members of a population, is it possible to have sampling errors when sampling from the population?
8. If participants are first paired before being randomly assigned to experimental and control groups, are the resulting data "independent" *or* "dependent"?
9. Which type of data tends to have less sampling error (circle one)?
   A. Independent.
   B. Dependent.

Question for Discussion

10. Describe three possible research questions with an independent (categorical data) variable and a dependent variable (continuous data) that would require the use of an Independent Samples t Test, a Dependent Samples t Test, or a One-Sample t Test (refer to Table 24.1 in Chapter 24).

Notes

1. As indicated in Chapter 21, when a result is statistically significant, the null hypothesis is rejected.
2. See Appendix A for the computational procedures for t tests for reference.
3. In the types of study we are considering, researchers do not know the variation in population. However, the t test uses the standard deviations of the samples to estimate the variation of the population. In other words, the standard deviations of the samples provide the t test with an indication of the amount of variation in the populations from which the samples were drawn.

4. Ideally, we would like to conduct an experiment in which the two groups are initially *identical* in their visual acuity.

5. Description of the computation formula, for additional conceptual understanding of how the *t* values are calculated, is discussed in Appendix A. However, it is helpful to observe that in all three types of *t* tests, the computation formulas share in common taking the difference between two means being compared, then dividing it by the standard error of the difference between means in the case of independent or related sample means, and standard error of mean in the case of a One-Sample *t* Test. The formulas for the standard error of the difference between means are different for Independent Samples *t* Test and dependent *t* test. These, along with the formula for the standard error of mean, are listed in Appendix A.

CHAPTER 25

Independent Samples *t* Test

Chapter Objectives

The reader will be able to:

- ❏ Explain the use of the Independent Samples *t* Test as a test of two independent sample means.
- ❏ Differentiate the role of independent and dependent variables.
- ❏ Explain the concept of standard error of the difference between means.
- ❏ Interpret the results through the probability level that the null hypothesis is true.

As mentioned earlier, the common use of *t* tests is to simply compare two means to see whether they are statistically different. And one of the common means comparisons is an ***Independent Samples t Test***. The purpose of an Independent Samples *t* Test is to compare the difference between two independent sample means.

As described in Example 1 in Chapter 24, an Independent Samples *t* Test has independent data, which means that there is no pairing or matching of individuals across the two samples. In the example, men and women were drawn independently from the two populations without regard to

The purpose of ***Independent Samples t Test*** is to compare the difference between two independent sample means.

DOI: 10.4324/9781003299356-31

In order to perform the Independent Samples *t* Test, we need one continuous variable to produce the group means and one categorical data to make up the groups.

whether each individual in one group "matches" in any way (such as age) an individual in another group. And the data are normally distributed.[1]

Let's take a closer look at this example. In determining whether there is a difference between a random sample of 50 men and 50 women voters in their attitude toward welfare, their attitude scores showed that women had a mean of 38 (on a scale from 0 to 50, where 50 was the most favorable attitude) while men had a mean of 35. Notice that in this example, the attitude scores is a continuous data, more specifically, interval data. And men and women represent categorical data. In order to perform the Independent Samples *t* Test, we need one continuous data to produce the group means and one categorical data to make up the groups (see Table 24.1).

The attitude scores depend on the men and women groups. Functionally, the attitude scores are the outcomes based on the groups, which become the *dependent variable* or the test variable. The ***dependent variable*** is a variable that *depends* on or is the outcome of the *independent variable* or the predictor. And the predictor or the independent variable is the categorical data that is used to predict the value of the dependent variable. The ***independent variable*** is the variable that may affect the dependent or the outcome variable. In other research scenarios, the independent variable may have two groups whose differences you may be interested in comparing (e.g., experimental group and control group, athletes and non-athletes, alcoholics and non-alcoholics, etc.).

A ***dependent variable*** is a variable that depends on or is the outcome of the independent variable or the predictor.

An ***independent variable*** is a variable that may affect the dependent or the outcome variable.

The question is: Are the mean scores in the dependent variable affected by which group one belongs to? There are two possible explanations for the mean difference in this example. One possibility is that while there is a difference of three points, this difference may not be statistically significant. In fact, this difference may be due to sampling errors created by random sampling. This would be the null hypothesis expressed in the following way.

Null hypothesis: H_0: $\mu_1 = \mu_2$ or H_0: $\mu_1 - \mu_2 = 0$

Alternatively, the other possible explanation for this mean difference is that there is a true *difference* between men and women in their attitude scores toward welfare.

Alternative hypothesis: H_a: $\mu_1 \neq \mu_2$ or H_a: $\mu_1 - \mu_2 \neq 0$

Notice that the notations used in these hypotheses are the equal (=) and the not equal (≠) signs, instead of the greater than (>) or less than (<) signs. This indicates a non-directional (two-tailed) hypothesis. We are interested in the possible statistical *difference* instead of determining whether the women's attitude scores are *greater* than the men's attitude scores. If we are able to reject the null hypothesis and establish a statistical *difference* between the two means through non-directional hypothesis testing, we would also know that the women's mean is statistically higher by observing the numbers without having to do a separate directional hypothesis testing.

A critical question to ask at this point is, if we were to repeat the study by measuring the attitude scores of men and women of a similar random sample, how close would these means be to the actual population mean if we had studied the entire population? It could be entirely possible that in a repeated study, the men's mean attitude score turns out to be higher than the women's. So, what would be the difference between the first set of means in the first study and the second set of means from a repeated study? In other words, what would be the average expected difference between the means? This is known as the ***standard error of the difference between means***. While the standard error of the difference between means and the standard error of the means are similar in that when the samples are larger, the standard errors become smaller, increasing the likelihood of finding statistical significance, standard error of the difference between means is different from standard error of the means (discussed in Chapter 8) in that it measures the average amount of error from the two sample means using the standard error of each sample.

The calculation of the standard error of the difference between means involves the standard error of each group as well as the group size. In addition, the formula for calculating the standard error of the difference between means needs to be further adjusted if the standard deviations between the groups are not considered to be equal. Depending on whether the group sizes between the groups can be considered equal or unequal, and whether the standard deviations can be considered equal or unequal, different adjustments need to be made to the calculation formula, which is beyond the scope of this book.[2] Nevertheless, it is important to understand the concept behind this important number because it is divided into the mean difference between the two groups in calculating the *t* value for the Independent Samples *t* Test as the formula suggests. Conceptually then, the standard error of the difference between means is the average expected difference between two random samples from two different populations on a measurement.

The sample size of the two independent samples is considered by calculating the degrees of freedom (*df*), which is calculated by adding the two samples and subtracting two. Thus, the degrees of freedom for

The ***standard error of the difference between means*** measures the average amount of error from the two sample means using the standard error of each sample.

Independent Samples t Test is the total number of participants minus the two parameters (means):[3]

$$df = n_1 + n_2 - 2$$

Another way to calculate the degrees of freedom is by totaling the degrees of freedom from the two samples:

$$df_1 + df_2 \text{ (where } df_1 = n_1 - 1 \text{ and } df_2 = n_2 - 1)$$

The formula in Appendix A shows the calculation for how the observed difference between the two means significantly differ relative to the standard error of the difference between means.

How do we know if the t value is significant? A t value can be calculated. And using the degrees of freedom and critical values of the t distribution table, also known as the t table, which one can find in any statistical textbook, one can determine the significance. Fortunately, statistical software programs such as SPSS calculate the t value as well as the probability of getting the t value that size or higher (p value). Simply based on the probability, we would determine whether the difference between the two independent sample means is statistically significant.[4]

When reporting the results of an Independent Samples t Test, one would typically report the means and standard deviations as well as the number of cases in a table form before stating the results (see Table 25.1).

In journals, there may be variations in how the results are reported. For example, one might see results reported as follows:

The null hypothesis was rejected at the 0.01 level [$t = 7.28$, $df = 98$, $p < 0.01$, two-tailed test].

The scientific notation above reports the t value, degrees of freedom, and indicates that the p value is less than 0.01. This leads to the conclusion that we can reject the null hypothesis. There is a significant difference between boys and girls in their height.

When reporting the results of an Independent Samples t Test, one would typically report the means and standard deviations as well as the number of cases on a table form before stating the results.

When the null hypothesis is rejected, we know that the difference between the means was statistically significant.

TABLE 25.1
Means and Standard Deviations of Attitudes Scores toward Welfare

	Mean	Standard Deviation	Sample Size n
Men	35	2.01	50
Women	38	2.11	50

When the null hypothesis is rejected, we know that the difference between the means was statistically significant. In this example, the difference between men and women in their attitude scores was statistically significant—that the difference of three points was not due to sampling error.

A Dependent Samples *t* Test is similar to an Independent Samples *t* Test, but it is used in a different scenario where the sample groups are related (e.g., pretest and posttest). This type of *t* test is described next in Chapter 26.

Exercise for Chapter 25

Factual Questions

1. What scenario is required when performing an Independent Samples *t* Test?
2. Assume that a researcher is interested in growth spurts among young teens decided to measure the height difference between boys and girls in middle school from a random sample of 200 boys and girls. In this study, what are the two variables of interest, and which of the two variables is the independent variable and which is the dependent variable?
3. In the above scenario, state the null and the alternative hypotheses.
4. If the alternative hypothesis is as follows: $H_a: \mu_1 \neq \mu_2$, and you rejected the null hypothesis in favor of the alternative hypothesis, state in words what you would conclude about the difference in height between boys and girls.
5. The above alternative hypothesis would be (circle one)?
   A. Non-directional hypothesis.
   B. Directional hypothesis.
6. In another study, if the alternative hypothesis is $\mu_1 < \mu_2$, and you concluded in favor of the alternative hypothesis, in words what you would conclude?
7. The above alternative hypothesis would be (circle one)?
   A. Non-directional hypothesis.
   B. Directional hypothesis.
8. Generally, the larger the samples, the smaller the standard errors become, increasing the likelihood of finding statistical significance. This statement is (circle one)?
   A. True.
   B. False.

9. Calculating the standard error of the difference between means depends on the standard deviations as well as the sample sizes between the groups, and can be considered equal or unequal. This statement is (circle one)?
   A. True.
   B. False.
10. What do the following results indicate in terms of rejecting the null hypothesis: ($t = 3.20$, $df = 28$, $p < 0.05$, two-tailed test)?

Question for Discussion

11. Give an example of one continuous data and one categorical data with two categories that would require performing an Independent Samples t Test.

Notes

1. When the normality of data is violated, Mann–Whitney U Test (for two levels of independent variable) or Kruskal–Wallis H Test (for more than two levels of independent variable) are conducted. This is beyond the scope of this book.
2. If the sample sizes or the standard deviations are very different between the two independent groups, one needs to consider non-parametric tests such as the Mann–Whitney U test as an alternative.
3. A full explanation of the degrees of freedom is beyond the scope of this book.
4. Review Chapter 22 on probability.

CHAPTER 26

Dependent Samples *t* Test

Chapter Objectives

The reader will be able to:

- ❏ Explain the use of a Dependent Samples *t* Test as a test of the difference between two related samples.
- ❏ Interpret the results through the probability level that the null hypothesis is true.

Similar to the Independent Samples *t* Test, a Dependent Samples *t* Test compares two means. However, these two means come from one sample with repeated measures or from two matched samples. And since the data is from one sample or related samples, we perform a **Dependent Samples *t* Test.** Thus, the purpose of a Dependent Samples *t* Test is to compare the difference between two related or dependent samples. For example, if students learning English as a second language took a placement test at the beginning of the school year and took the test again at the end of the school year, this would be an example of one sample with two repeated measures, the same students measured twice over time. This means that for each student there are two test scores. This kind of before-and-after comparison is also

> The purpose of a **Dependent Samples *t* Test** is to compare the difference between two related or dependent samples.

DOI: 10.4324/9781003299356-32

A before-and-after comparison is also known as a ***within-subject design or repeated-measures design***.

known as a ***within-subject design or repeated-measures design***, which is very common in education-related contexts where knowledge gained over time is measured. The advantage of this design is that since there is just one sample, this controls for any pre-existing individual differences between samples. In addition, having one sample is economical. However, participants' repeated tasks such as repeated tests may show an improvement in scores just from practice having seen the task before.

In another example, let us further explore Example 2 from Chapter 24. In this example, pairs of same-sex siblings (two brothers or two sisters) comprised the samples. They are pairs of siblings. Through a coin toss, they were put into one of either group: one that received a vitamin supplement and the other that received a placebo. For each person in the experimental group, there was a same-sex sibling in the control group. So, these samples are two matched samples. In measuring the effects of the vitamin supplement intake, it makes sense to study same-sex siblings who are similar in their family diet and habits, similar family setting, and so on.

A *Dependent Samples t Test* compares means just like the Independent Samples *t* Test.

Different from independent samples, in a Dependent Samples *t* Test, the matching or paired participants in the dependent groups have already built in a similarity between the groups or repeated data points, making it possible to reduce error.

In both examples, the Dependent Samples *t* Test procedure requires one continuous data and one categorical data with two categories. This is no different from the Independent Samples *t* Test, as well as the assumption of normality of data, but the Dependent Samples *t* Test assumes that the samples are related observations are independent.[1] (See Table 24.1 in Chapter 24.)

In another example, suppose when the children's energy level was compared between the experimental group and the control group, the experimental group showed on average an energy level of 7.4 on a scale of 0 to 10, with 10 being the highest level of energy. The control group on the other hand showed an average of 5.55. The null hypothesis states that the difference between the two groups is equal to zero (0) or that there is no significant difference between the two groups in their energy level.

$$\text{Null hypothesis: } H_0\text{: } \mu_1 = \mu_2 \qquad \text{or} \qquad H_0\text{: } \mu_1 - \mu_2 = 0$$

Alternatively, the other possible explanation for this mean difference is that there is a true *difference* in mean energy levels between the siblings assigned to either one of the groups.

$$\text{Alternative hypothesis: } H_a\text{: } \mu_1 \neq \mu_2 \qquad \text{or} \qquad H_a\text{: } \mu_1 - \mu_2 \neq 0$$

The degree of freedom *(df)* in the Dependent Samples *t* Test is *N*, which represents the total number of pairs, minus 1:

$$df = N - 1$$

The degrees of freedom in the Dependent Samples *t* Test is *N* - 1.

In calculating for the *t* value, we take the difference between the two related samples. In the case of one sample with repeated measures, assuming that the "after" scores are higher than the "before" scores, we take the difference between the two means and divide this difference by the standard error of the difference between the means. (See Appendix A for the formula.)

In the Dependent Samples *t* Test, the standard error of the difference between means is calculated differently from the Independent Samples *t* Test, however, the same concept applies. The standard error of the difference between means is the average expected difference between dependent sample means.

Table 26.1 shows the results of the comparison of the energy levels between siblings.

According to the results, the probability (*p*) that the *t* value of 5.965 occurred by chance or due to sampling error is 0.000. Therefore, it can be concluded that the null hypothesis is rejected. There is a significant difference between the siblings in their energy levels. In a journal, one might state the following:

> A paired samples analysis produced a significant *t* value [*t* (19) = 5.965, *p* < 0.001]. There was a significant difference in the energy level for the experimental group (Mean = 7.40, SD = 1.095) compared with the control group (Mean = 5.55, SD = 1.099). Siblings in the experimental group that took the vitamin supplement showed a significantly *different* energy level than the others in the control group who did not take the vitamin supplements.

TABLE 26.1
Results of Dependent Samples *t* Test Comparing the Experimental and Control Group

	Number of Pairs	Cor.	Sig	Mean	SD	SE of Mean
Experimental	20	0.201	0.395	7.40	1.095	0.245
Control				5.55	1.099	0.246

Paired difference

Mean Difference	SD	SE of Mean	*t* Value	*df*	Sig	
1.85	1.39	0.310	5.965	19	0.000	

The scientific notation used in the statement indicates that a *t* test was performed. Then *df* in parentheses and *t* value are reported, followed by a report that the *p* value associated with the study was less than 0.001. Notice that the statement says that there was a significant *difference*, and not a *"significantly higher"* energy level. This is because we performed a non-directional hypothesis test. But just the same, since the test established a significant difference, we can safely assume that the experimental group showed a *significantly higher* mean.

Exercise for Chapter 26

Factual Questions

1. How is the Dependent Samples *t* Test different from the Independent Samples *t* Test in terms of samples?
2. What are the similarities in the assumptions behind both the dependent and Independent Samples *t* Tests?
3. What is a Dependent Samples *t* Test with one sample also known as?
4. In a Dependent Samples *t* Test, what is the *df* for a sample of 100 participants in a particular study?
5. The Dependent Samples *t* Test would have less sampling error in its design. Discuss the reasons.
6. Given the results below what would you conclude about the mean difference between Group A and Group B?

Group	Number of Pairs	Cor.	*p*	Mean	SD	SE of Mean
Group A	50	0.087	0.549	7.64	1.03	0.145
Group B				4.82	1.32	0.187

Paired difference

Mean Difference	SD	SE of Mean	*t* Value	*df*	*p*	
2.82	1.60	0.226	12.46	49	0.000	

Question for Discussion

7. Give an example of one continuous data and one categorical data with two categories that would require performing a Dependent Samples *t* Test.

Note

1. When the normality of data is violated, the Wilcoxon matched-pairs signed ranks test (for two levels of dependent variable) or Friedman test (for more than two levels of dependent variable) are conducted. This is beyond the scope of this book.

CHAPTER 27

One-Sample *t* Test

Chapter Objectives

The reader will be able to:

- ❏ Explain the use of a One-Sample *t* Test as a test of the difference between a sample mean and a population mean.
- ❏ Interpret the results through the probability level that the null hypothesis is true.

Different from an Independent Samples *t* Test and a Dependent Samples *t* Test, **One-Sample *t* Test** is used when a sample mean is compared to a population mean or a historical mean. Recall Example 3 given in Chapter 24. Mathematics Assessment Test scores of the National Assessment of Educational Progress (NAEP) from 250 randomly selected students at a particular school were compared to the national mean score. The national mean represents the population mean and the mean from a group of students at a particular school represents the sample mean. These two means are compared to see whether the sample mean differs from the population mean. If the comparison shows no significant difference and the null hypothesis is true, one would conclude that the sample represents the population, that they are all part of the same population. If

A **One-Sample *t* Test** is used when a sample mean is compared to a population mean or a historical mean.

DOI: 10.4324/9781003299356-33

The degree of
freedom for a One-
Sample *t* Test is $N - 1$.

the comparison shows a significant difference and the null hypothesis is rejected, one would conclude that the sample is different from the population. The degree of freedom for a One-Sample *t* Test is $N - 1$.

Let's consider another example. Assume that anxiety scores as measured by an anxiety assessment inventory are normally distributed with the population mean of 22. A random sample of 40 patients who underwent treatment for anxiety showed a mean anxiety score of 20 with a standard deviation of 4. The question is: Is there a significant difference between the mean of the patients under treatment from the population mean?

Here are the important numbers in order to conduct a One-Sample *t* Test.

Population mean = 22

Sample mean = 20, SD = 4, *n* = 40

The null hypothesis would state that there is no difference between the sample mean and the population mean—that the patients who underwent treatment did not show any difference from the population:

Null hypothesis H$_0$: $\bar{x} = \mu$ or $\mu_1 - \mu_2 = 0$

The alternative hypothesis would state that there is a statistical difference between the sample mean and the population mean—that the patients who underwent treatment showed a difference from the population. This would be a non-directional, two-tailed hypothesis.

Alternative hypothesis H$_a$: $\bar{x} \neq \mu$ or $\mu_1 - \mu_2 \neq 0$

Another possible alternative hypothesis would be the directional, one-tailed hypothesis that states that one mean is greater than the other. While the null hypothesis still remains the same, the directional hypothesis would take on the symbols < or > when expressing the alternative hypothesis. As explained in Chapter 24, because of redundancy in these two possibilities of the alternative hypothesis, we will proceed with the non-directional, two-tailed hypothesis.

As is true in all parametric tests, there are several assumptions with the One-Sample *t* t Test. The measurement has to be continuous data that is normally distributed.

The *t* value involves taking the difference between the sample mean and the population mean relative to the standard error of mean. See Appendix A for the formula to calculate the *t* value.

The **standard error of mean** is calculated by dividing the standard deviation by the square root of the sample size N. For the definition of the standard error of mean, review Chapter 8.

The computation formula and calculation are listed in Appendix A for your reference. In addition, the *t* value, as well as the probability value that the null hypothesis is true, can be produced from SPSS or similar software programs. Based on the probability value, the results of a One-Sample *t* Test might state:

> **There is a significant difference between the sample mean and the population mean [t (39) = 3.16, $p < 0.01$].**

This means that the sample mean does not represent the population mean. This might imply that the anxiety treatment worked to significantly change the anxiety level of the participants in the random sample.

Exercise for Chapter 27

Factual Questions

1. Which of the following statements is true (circle one)?
   A. A One-Sample *t* Test is used to test the difference between a sample mean and a population mean to determine statistical significance.
   B. A One-Sample *t* Test is used to test the difference between two sample means from one sample.
2. SAT verbal scores are normally distributed with the population mean of 500. A local high school has instituted a new program to engage students in reading. A sample of 90 students from this high school is randomly selected following their participation in this reading program and their SAT verbal score mean of 520 was compared to the national mean. Discuss how this would require a One-Sample *t* Test.
3. For the above example, how would you set up a two-tailed alternative hypothesis, and what would you conclude if the null hypothesis were rejected?
4. How would you set up a one-tailed alternative hypothesis for the above example and what would you conclude if the null hypothesis was rejected?

5. Assume that a One-Sample t Test showed the following results: [t (49) = 4.21, $p < 0.01$]. Answer the following questions.
   A. What is the probability that the null hypothesis is true?
   B. What is the df of the study?
   C. What is the n of the study?
   D. What would you conclude?

Question for Discussion

6. Discuss a possible study topic that would require a One-Sample t Test.

CHAPTER 28

Reporting the Results of *t* Tests

Display of Outcomes

Chapter Objectives

The reader will be able to:

❏ Apply how the results of *t* tests are reported in various widely used forms such as a table or statement using various phrases and wording.

❏ Distinguish the difference between statistically significant and practical significance.

In Chapters 24–27, the use of the *t* test to test the difference between two sample means for significance was considered. Obviously, the values of the means should be reported before the results of the statistical test performed on them are reported. In addition, the values of the standard deviations and the number of cases in each group should be reported first. This may be done within the context of a sentence or in a table. Table 28.1 shows an example of a mean and standard deviation table.

In this example, the samples for Groups A and B were drawn at random. The null hypothesis states that the 3.50 point difference (6.00 − 2.50 = 3.50) between the means of 2.50 and 6.00 is the result of sampling errors (i.e., errors resulting from random sampling) and that the true difference

> The means, standard deviations, and numbers of cases should be reported before the results of a *t* test are reported.

DOI: 10.4324/9781003299356-34

TABLE 28.1

Means and Standard Deviations

	M	SD	n
Group A	2.50	1.87	6
Group B	6.00	1.89	6

in the population is zero. (Refer back to Chapters 11 and 12 on the concept of mean and Chapter 14 on the concept of standard deviation.)

The result of a significant *t* test may be described in several ways. Below are some examples of the results of this example. The statement in Example 1 below implies that the null hypothesis has been rejected. Note that the term *statistically significant* is synonymous with *rejecting the null hypothesis.*

<table>
<tr><td align="center">EXAMPLE 1</td></tr>
<tr><td>The difference between the means is statistically significant ($t = 3.22$, $df = 10$, $p < 0.01$).</td></tr>
</table>

<table>
<tr><td align="center">EXAMPLE 2</td></tr>
<tr><td>The difference between the means is significant at the 0.01 level ($t = 3.22$, $df = 10$).</td></tr>
</table>

In Example 2, the researcher used slightly different wording to indicate that significance was obtained at the 0.01 level. The phrase *significant at the 0.01 level* indicates that *p* was equal to or less than 0.01, which is the probability that the null hypothesis is correct. Thus, the null hypothesis was rejected.

Example 3 provides the same information as Examples 1 and 2 but with different wording. The sentence indicates that the difference is statistically significant because *rejecting the null hypothesis* is the same as *declaring statistical significance.*

Statistically significant is synonymous with *rejecting the null hypothesis.*

This is one way to report the result of a significant *t* test.

The phrase *significant at the 0.01 level* indicates that *p* was equal to or less than 0.01.

Rejecting the null hypothesis is the same as *declaring statistical significance.*

EXAMPLE 3

The null hypothesis was rejected at the 0.01 level ($t = 3.22$, $df = 10$).

Any of the forms of expression illustrated in the previous three examples are acceptable. However, authors of journal articles seldom explicitly mention the null hypothesis. Instead, they tend to use the forms of expression in Examples 1 and 2. In theses and dissertations, in contrast, explicit references to the null hypothesis are more common.

When researchers use the word *significant* in reporting the results of significance tests, they should modify it with the adjective *statistically*. This is because a result may be *statistically significant* but not of any *practical significance*. For instance, suppose a researcher found a statistically significant difference of two points in favor of a computer-assisted approach over a traditional lecture/textbook approach to teaching. While it may be statistically significant, it may not be of practical significance if the school district has to invest sizeable amounts of money to buy new hardware and software. In other words, the cost of the difference may be too great in light of the absolute size of the benefit.[1]

A result may be statistically significant but not of any practical significance.

Now, consider how researchers report the results of a *t* test when the difference between means is not statistically significant. Table 28.2 presents descriptive statistics of means and standard deviations. Examples 4 through 6 show some ways to express the results of the insignificant *t* test for the data in the table.

TABLE 28.2
Means and Standard Deviations

	M	SD	*n*
Group A	8.14	2.19	7
Group B	5.71	2.81	7

EXAMPLE 4

The difference between the means is not statistically significant ($t = 1.80$, $df = 12$, $p > 0.05$).

EXAMPLE 5
For the difference between the means, $t = 1.80$ ($df = 12$, *n.s.*).

The fact that p is *greater than* (>) 0.05 in Example 4 indicates that the null hypothesis was not rejected.

The author of Example 5 has used the abbreviation *n.s.* to indicate that the difference is *not significant.* Because the example does not indicate a specific probability level, most readers will assume that it was not significant at the 0.05 level—the most liberal of the widely used levels.[2] Example 4 is preferable to Example 5 because the former indicates the specific probability level that was used to test the null hypothesis.

EXAMPLE 6
The null hypothesis for the difference between the means was not rejected at the 0.05 level ($t = 1.80$, $df = 12$).

Example 6 shows how the results of the test might be expressed with explicit reference to the null hypothesis.

While reading journal articles, theses, and dissertations, you will find variations of words used to report the results of a t test. The examples in this chapter illustrate some of the most widely used forms of reporting.

Exercise for Chapter 28

Factual Questions

1. Which statistics should be reported before the results of a t test are reported?
2. Suppose you read this statement: "The difference between the means is statistically significant at the 0.05 level ($t = 2.333$, $df = 11$)." Should you conclude that the null hypothesis has been rejected?

3. Suppose you read this statement: "The null hypothesis was rejected ($t = 2.810$, $df = 40$, $p < 0.01$)." Should you conclude that the difference is statistically significant?

4. Suppose you read this statement: "The null hypothesis was not rejected ($t = -0.926$, $df = 24$, $p > 0.05$)." Describe in words the meaning of the statistical term "$p > 0.05$."

5. For the statement in Question 4, should you conclude that the difference is statistically significant?

6. Suppose you read this statement: "For the difference between the means, $t = 2.111$ ($df = 5$, *n.s.*)." Should you conclude that the null hypothesis has been rejected?

7. Which type of author seldom explicitly mentions the null hypothesis?
   A. Authors of dissertations.
   B. Authors of journal articles.

Question for Discussion

8. What type of *t* test was done in the following studies: Independent Samples *t* Test, Dependent Samples *t* Test, or One-Sample *t* Test?
   A. There was a significant increase in alcohol consumption among college students one week after the end of the semester compared with one week before the end of the semester, $t(122) = 3.5$, $p < 0.01$).
   B. Students who underwent a math intervention program showed a mean test score that was significantly different from students who did not participate, $t(40) = 2.0$, $p < 0.05$).
   C. College students who drank coffee regularly were more awake in class than the college average, $t(200) = 3.1$, $p < 0.05$).

Notes

1. Practical significance is considered in greater depth again in Chapter 32.
2. The 0.05 level is the "most liberal" in the sense that it is the level most likely to permit rejection of the null hypothesis. In other words, if a researcher uses the 0.01 or 0.001 levels, he or she is less likely to reject the null hypothesis than if the 0.05 level is used.

CHAPTER 29

One-Way ANOVA

Chapter Objectives

The reader will be able to:

- ❏ Recognize that analysis of variance is used to test the difference(s) among two or more means.
- ❏ Examine how analysis of variance is reported.
- ❏ Explain the concept of One-Way ANOVA and its purpose and use of post-hoc tests.

The t test, which tests the null hypothesis regarding the difference between *two* means, was covered in Chapters 24 through 27. A closely related statistical procedure is ***analysis of variance*** (*ANOVA*), which produces what is called an F test. An F test can be used to test the difference(s) among *two or more* means.

Like the t test, ANOVA can be used to test the difference between two means. When this is done, the resulting *probability* will be the same as the probability that would have been obtained with a t test. However, the value of F will not be the same as the value of t.[1]

Let's say that a researcher tests the effects of the amount of tutoring, as measured in three levels (independent variable) on students'

*Analysis of variance (**ANOVA**) produces an F test to measure the difference(s) among two or more means.*

DOI: 10.4324/9781003299356-35

achievement test scores (dependent variable). We would need to conduct an ANOVA that yields an F test to answer the question: "Are there differences among the means of test scores in the three different levels of tutoring?" The F test, when significant, will show that there is a significant effect of tutoring on test scores overall. This means that the different levels of tutoring will change the test scores significantly. This is an important finding, but general in that we don't yet know the significances among the three levels in their varying effect on the test scores.

ANOVA can also be used to test the differences among more than two means in a single test, which cannot be done with a t test.

Consider an example with reported means in Example 1.

EXAMPLE 1

A new drug for treating migraine headaches was tested on three samples selected at random from a population. The first group received 250 milligrams, the second received 100 milligrams, and the third received a placebo (an inert substance). The average pain reported for the three groups (on a scale from 0 to 20, with 20 representing the most pain) was determined by calculating the means. The means for the groups were:

Group 1: $M = 1.78$
Group 2: $M = 3.98$
Group 3: $M = 12.88$

As you can see, there are three differences in the means comparison possible among the three groups:

1. The difference between Groups 1 and 2 (1.78 versus 3.98).
2. The difference between Groups 1 and 3 (1.78 versus 12.88).
3. The difference between Groups 2 and 3 (3.98 versus 12.88).

The *set of three differences* in Example 2 can be tested with a single ANOVA.

Instead of running three separate t tests,[2] a researcher can run a single F test using ANOVA to test the significance of this *set of three differences*. Examples 2 and 3 are two ways the results of an F test could be reported.

EXAMPLE 2

The difference among the means of three groups was statistically significant at the 0.01 level ($F = 58.769$, $df = 2, 36$).

Note that this statement in Example 2 is similar in structure to the results for a *t* test (see Chapter 28). The statement indicates that there is a significant difference. Thus, the null hypothesis may be rejected at the 0.01 level. The null hypothesis for this test says that the *differences* among the three means was created at random. By rejecting the null hypothesis, a researcher is rejecting the notion that *one or more* of the differences were created at random by sampling error. Said another way, there are one or more significant differences among these comparisons. Notice that the test does not indicate which of the three differences is responsible for the rejection of the null hypothesis but that *at least one of the three comparisons* is statistically significant.[3]

Table 29.1 shows a second way that the results of F tests conducted with ANOVA are commonly reported in journals. It is called an ***ANOVA table***. The table shows the same values of F, *df*, and p that are reported in Example 1. The table also shows the values of the *sum of squares* and *mean square*, which are intermediate values obtained in the calculation of F. (For example, if you divide the mean square of 315.592 by the mean square of 5.370, you will obtain F.) For the typical consumer of research, the values of the sum of squares and mean square are of little interest. A consumer is primarily interested in whether the null hypothesis can be rejected, which is indicated by the value of p.

Often, the value of p will be given in an ANOVA table as well as in the text of the research report. For instance, a researcher might include a statement like the one in Example 2 in the text.

Another way to look at the result is by measuring the ***effect size***. Effect size is defined as measuring the percentage of the unknown in one

The method of reporting the results of an ANOVA is sometimes similar to the method for a *t* test.

When a researcher rejects the null hypothesis with ANOVA, the researcher is rejecting the notion that *one or more* of the differences in the set were created at random.

An **ANOVA table** is sometimes used to report the results.

The *sum of squares* and *mean square* in an ANOVA table are of little interest to the typical consumer of research.

Effect size is defined as measuring the percentage of the unknown in one variable that is explained by another variable.

TABLE 29.1
Analysis of Variance Table for the Data in Example 1

Source of Variation	*df*	Sum of Squares	Mean Square	F	p
Between groups	2	631.185	315.592	58.769[4]	0.009
Within groups	36	193.320	5.370		
Total	38	824.505			

Eta squared (η^2) measures how much of the variable is explained by group differences.

variable that is explained by another variable. Effect size in ANOVA is measured through ***eta squared (η^2)***. *Eta squared* measures how much of the variable is explained by group differences. The calculation is simple— it is between groups sums of squares divided by total sums of squares. Using the results in Table 29.1, it is 631.185 divided by 824.505, which equals 0.7655 or 76.55%. This means that 76.55% of the variation in the variable is explained by difference between groups. More details about effect size for various contexts and what it means will be discussed in Chapter 32.

ANOVA can be used to compare the means of many groups. Consider Example 3. A single ANOVA can determine whether the null hypothesis for the entire set of six differences should be rejected. If the result is not significant, the researcher is done. There is no mean difference. If the result is significant, this only indicates that there is at least one or more significant means comparison, but it doesn't indicate which pair(s). In order to see which specific group pairs show significant difference, the researcher may conduct additional tests to determine which specific difference(s) is significant. A number of different tests, which do not always lead to the same conclusions, are available. These are known as ***post-hoc tests*** (after-the-fact tests). Among those that you may encounter are ***Tukey's HSD test and Scheffé's test***. The basic idea behind all post-hoc tests is the same in that each group mean is compared to each of the other group means. While these and other additional post-hoc tests are beyond the scope of this book, you will be able to understand them because they all result in a probability level (p), which is used to determine significance.

A ***post-hoc test*** (***Tukey's HSD test*** and ***Scheffé's test***) determines which pairs of means are significantly different.

In a ***One-Way ANOVA***, participants are classified in only one way.

The examples we have been considering are examples of what is known as a ***One-Way ANOVA*** (*also known as a **single-factor ANOVA***) in which we have a single factor with multiple categories or groups. This term is derived from the fact that participants are classified in only *one* way. In Examples 1 and 2, the participants were classified according to three groups to which they were assigned. In Example 3, they were classified according to four methods of instruction to which they were exposed.

EXAMPLE 3

Four methods of teaching computer literacy were used in an experiment, which resulted in four means. This produced these six differences:

The difference between Methods 1 and 2.
The difference between Methods 1 and 3.
The difference between Methods 1 and 4.
The difference between Methods 2 and 3.
The difference between Methods 2 and 4.
The difference between Methods 3 and 4.

Chapter 30 introduces and illustrates a ***Two-Way ANOVA*** (also known as ***Two-factor ANOVA***), in which each participant is classified in two ways, such as (1) which drug group they were assigned to and (2) whether each participant is male or female, each with multiple categories. Two-Way ANOVA involves two categorical grouping variables and a continuous dependent variable as summarized in Table 29.2.

Two-Way ANOVA (also known as *Two-factor ANOVA*) classifies participants in two ways to examine two main effects and one interaction effect.

TABLE 29.2
Summary of One-Way and Two-Way ANOVA Designs

	Grouping Variable 1	Grouping Variable 2	Test Variable
	(Independent Variable)		(Dependent Variable)
One-Way ANOVA	Two or more levels or categories	N/A	A continuous data
Two-Way ANOVA	Two or more levels or categories	Two or more levels or categories	A continuous data
Example	Three tutoring levels	Three levels of attitude toward academic achievement	Test scores

The Two-Way ANOVA permits researchers to answer interesting questions such as: Are some drugs more effective for treating men than they are for treating women? Consider Table 29.2 for a better distinction between One-and Two-Way ANOVAs. More on Two-Way ANOVA will be discussed next in Chapter 30.

Exercise for Chapter 29

Factual Questions

1. ANOVA stands for what three words?
2. What is the name of the test that can be conducted with an ANOVA?
3. "An ANOVA can be appropriately used to test *only* the difference between two means." Is this statement "true" *or* "false"?
4. If the difference between a pair of means is tested with ANOVA, will the probability level be different from that where the difference was tested with a *t* test?
5. Which statistic in an ANOVA table is of greatest interest to the typical consumer of research?
6. Suppose you read this statement: "The difference between the means was not statistically significant at the 0.05 level ($F = 2.293$, $df = 12$, 18)." Should you conclude that the null hypothesis was rejected?
7. Suppose you read this statement: "The difference between the means was statistically significant at the 0.01 level ($F = 3.409$, $df = 14$, 17)." Should you conclude that the null hypothesis was rejected?
8. Suppose you saw this in the footnote to a One-Way ANOVA table: "$p < 0.05$." Are the differences statistically significant?
9. Suppose participants were classified according to their grade level in order to test the differences among the means for the grade levels. Does this call for a "One-Way ANOVA" *or* a "Two-Way ANOVA"?
10. Suppose that the participants were classified according to their grade levels and their country of birth in order to study differences among means for both grade level and country of birth. Does this call for a "One-Way ANOVA" *or* a "Two-Way ANOVA"?

Question for Discussion

11. Briefly describe a hypothetical study in which it would be appropriate to conduct a One-Way ANOVA but *not* appropriate to conduct a *t* test.

Notes

1. Historically, the t test preceded ANOVA. Because ANOVA will also test the difference between two means, the t test is no longer needed. However, for instructional purposes, the t test is still taught in introductory statistics classes and it is still widely used by researchers when only two means are being compared.
2. It would be inappropriate to run three separate t tests without adjusting the probabilities for interpreting t. Such adjustments are not straightforward. However, a single F test automatically makes appropriate adjustments to the probabilities.
3. Procedures for determining which individual differences are significant are beyond the scope of this book.
4. $p < 0.01$.

CHAPTER 30

Two-Way ANOVA

Chapter Objectives

The reader will be able to:

- ❏ Describe the use and examples of Two-Way ANOVA.
- ❏ Explain that Two-Way ANOVA has two main effects and one interaction effect.
- ❏ Interpret the main effects and interaction effect.

In a **Two-Way ANOVA** (also known as a *two-factor ANOVA*), participants are classified in two ways. Consider Example 1, which illustrates this.

EXAMPLE 1
A random sample of welfare recipients was assigned to a new job-training program. Another random sample (with the same sample size) was assigned to a conventional job-training program. (Note: the type of job-training program they were assigned to is one of the ways

In **Two-Way ANOVA**, participants are classified in two ways.

DOI: 10.4324/9781003299356-36

in which the participants were classified.) Participants were also classified according to whether or not they had a high school diploma. All of the participants in each group found employment in the private sector at the end of their training. Their mean hourly wages are shown in the following table.[1]

Education	Types of Job-Training Program		Row Mean
	Conventional Program	New Program	
HS diploma	M = $18.88	M = $18.75	M = $18.82
No HS diploma	M = $14.56	M = $18.80	M = $16.68
Column Mean	M = $16.72	M = $18.78	

First, consider the Column Means of $16.72 (for the conventional program) and $18.78 (for the new program). These suggest that, overall, the new program is superior to the conventional one. In other words, if we temporarily ignore whether participants have a high school diploma or not, the new program seems superior to the conventional one. This difference ($18.78 – $16.72 = $2.06) suggests that there is what is called a ***main effect***. A *main effect* is the result of comparing one of the ways in which the participants were classified while temporarily ignoring the other way in which they were classified.

Because the concept of a *main effect* can be difficult to grasp at first, let's consider it again. You can see that the Column Mean of $16.72 is for all of those who had the conventional program regardless of whether they have a high school diploma. The Column Mean of $18.78 is for all those who had the new program—some regardless of education. Thus, by looking at the Column Means, only the effect of the type of program is being considered (and *not* the effect of a high school diploma). When researchers examine the effect of only one of the ways in which the participants were classified, they are examining a *main effect*. Thus, the difference between Column Means measures the *main effect*.

Similarly, let's consider the Row Means of $18.82 (for those with a high school diploma) and $16.68 (for those without a high school diploma). This suggests that those with a diploma, on average, have higher earnings than those without one. This is also a *main effect*. This main effect is high school diploma while temporarily ignoring the type of training program.

A ***main effect*** results from examining one way in which participants were classified while ignoring the other.

If there is a difference between row means, this suggests a *main effect*.

If there is a difference between column means, it suggests another main effect.

In other words, regardless of the type of program, those with a high school diploma earn more. This is the main effect of education.

Up to this point, there are two findings that would be of interest to those studying welfare: (1) the new program seems to be superior to the conventional program in terms of hourly wages, and (2) those with a high school diploma seem to have higher hourly wages. (Note: the term "seems to" is being used. We do not yet know whether the differences are statistically significant in light of sampling error.)

You may already have noticed that there is a third interesting finding: those with a high school diploma seem to earn about the same amount regardless of the program. This statement is based on these means for *those with a high school diploma* reproduced below from the first row of the table in Example 1.

	Conventional Program	New Program
HS diploma	$M = \$18.88$	$M = \$18.75$

In contrast, those without a high school diploma seem to benefit more from the new program than from the conventional one. This statement is based on these means for *those without a high school diploma* reproduced below from the second row of the table in Example 1.

	Conventional Program	New Program
No HS diploma	$M = \$14.56$	$M = \$18.80$

Now, suppose you were the researcher who conducted this study. You are now an expert on the training programs for those on welfare, and an administrator calls you for advice. She asks you, "Which program should we use? The conventional one or the new one?" You could, of course, tell her that there is a *main effect* for programs that suggests that, overall, the new program is superior in terms of wages upon employment. However, if this is all you told the administrator, your answer would be incomplete. A more complete answer would consist of two parts:

1. For those with a diploma, the two programs are about equal in their effectiveness. Other things being equal, it is not important which program is used with welfare recipients who have a high school diploma.
2. For those without a diploma, however, the new program is superior to the conventional one. Other things being equal, those without a diploma should be assigned to the new program, not the conventional program.

An **interaction** effect occurs when the effect of one grouping variable on the dependent variable is affected by the other grouping variable.

Because you cannot give a complete answer about the two types of programs (which is only one of the two ways in which the participants were classified) without also referring to high school diplomas (the other way in which they were classified), we say there is an *interaction* between the two variables (between the type of program and whether participants have a diploma). In other words, how well the two programs work depends in part upon whether or not the participants have a high school diploma. Thus, an interaction effect occurs when the effect of one grouping variable on the dependent variable is affected by the other grouping variable.

You can spot an *interaction* by subtracting and comparing the differences.

Here is a simple way in which you can spot an interaction when there are only two rows of means: subtract each mean in the second row from the mean in the first row. If the two differences are the same, there is no interaction. If they are different, there is an interaction. Here is how it works for the data in Example 1: $18.88 − $14.56 = **$4.32** and $18.75 − $18.80 = −**$0.05**. Because **$4.32** is *not* the same as −**$0.05**, the data suggest that there is an interaction effect.

Education	Types of Job-Training Program	
	Conventional Program	**New Program**
HS diploma	$M = 18.88	$M = 18.75
No HS diploma	$M = 14.56	$M = 18.80
Difference	$4.32	−$0.05

Consider Example 2, in which there are no main effects but an interaction effect.

EXAMPLE 2

A random sample from a population of those suffering from a chronic illness was administered a new drug. Another random sample from the same population was administered a standard drug. Participants were also classified as to whether they were male or female. At the end of the study, improvement was measured on a scale from 0 (for no improvement) to 10 (for complete recovery). These means were obtained:

	Standard Drug	New Drug	Row Mean
Male	$M = 5$	$M = 7$	$M = 6$
Female	$M = 7$	$M = 5$	$M = 6$
Column Mean	$M = 6$	$M = 6$	

Example 2 has no *main effects* but has an *interaction effect.* The two Column Means in Example 2 are the same. Thus, if we temporarily ignore whether participants are male or female, we would conclude that the two drugs are equally effective. To state it statistically, we would say that *there is no main effect for the drugs.*

The two Row Means are the same also. Thus, if we temporarily ignore which drug was taken, we can conclude that males and females improved to the same extent. To state it statistically, we would say that *there is no main effect for gender.*

Of course, the interesting finding in Example 2 is the *interaction.* The new drug works better for males and the standard drug works better for females. Subtracting as we did to identify an interaction for Example 1, we obtain the differences for Example 2, which are shown below. Because −2.00 is not equal to 2.00, there is an interaction effect.

Sometimes, the *interaction* is the most interesting finding.

	Standard Drug	New Drug
Male	$M = 5$	$M = 7$
Female	$M = 7$	$M = 5$
Difference	$M = -2$	$M = 2$

Consider Example 3, in which there are two main effects but no interaction.

EXAMPLE 3

Random samples of high and low achievers were assigned to one of two types of reinforcement during math lessons. Achievement on a math test at the end of the experiment was the outcome variable. The mean scores on the test were:

	Type of Reinforcement		
	Type A	Type B	Row Mean
High achievers	$M = 50$	$M = 30$	$M = 40$
Low achievers	$M = 40$	$M = 20$	$M = 30$
Column Mean	$M = 45$	$M = 25$	

In Example 3, there seems to be a main effect for the Type of Reinforcement as indicated by the difference between the Column Means (45.00 and 25.00). Thus, ignoring Achievement Levels temporarily, Type A seems to be more effective than Type B.

There also seems to be a main effect for achievement level as indicated by the difference between the Row Means (40.00 and 30.00). Thus, ignoring the type of reinforcement, high achievers score higher on the math test than low achievers.

However, there is no interaction, as indicated by the differences, which are shown below:

	Type of Reinforcement	
	Type A	Type B
High achievers	$M = 50$	$M = 30$
Low achievers	$M = 40$	$M = 20$
Difference	$M = 10$	$M = 10$

A Two-Way ANOVA will test the null hypotheses for the two *main effects* and the *interaction*.

The results of a Two-Way ANOVA are usually organized in a table.

What does this lack of interaction effect. indicate? It indicates that regardless of the type of reinforcement, high achievers are the same number of points higher than low achievers (i.e., 10 points). Put another way, regardless of whether students are high or low achievers, Type A reinforcement shows higher math test scores on average.[2]

In review, a Two-Way ANOVA examines two *main effects* and one *interaction effect*. Of course, because only random samples have been studied, null hypotheses should be tested. For each of the main effects and for the interaction, the null hypothesis states that there is no *true* difference—that the observed differences were created by random sampling errors. A Two-Way ANOVA will, therefore, test the two main effects and the interaction for significance. This is done by conducting three *F* tests (one for each of the three null hypotheses) and determining the probability associated with each. Typically, if a probability is 0.05 or less, the null hypothesis is rejected and the main effect or interaction effect being tested is declared statistically significant.

The results of a Two-Way ANOVA are usually organized in a table. While the entries in such tables sometimes vary,[3] the most important entries are shown in Example 4.

EXAMPLE 4		
Source	*F*	*p*
Achievement level	3.25	0.042*
Type of reinforcement	19.69	0.001**
Interaction (Ach. × Reinf.)	1.32	0.210

*$p < 0.05$, **$p < 0.01$

The probabilities shown in Example 4 indicate whether there is significance. Both main effects (i.e., Achievement level and Type of reinforcement) are statistically significant because the values of p are both less than 0.05, the most commonly used level for rejecting the null hypothesis. The interaction, however, is not significant because p is greater than 0.05, and the null hypothesis regarding this interaction should not be rejected. Thus, three null hypotheses have been tested with a Two-Way ANOVA, and two of the three null hypotheses have been rejected.

The probabilities (p) indicate whether there is significance.

Exercise for Chapter 30

Factual Questions

(Assume there are equal numbers of participants in each cell.)

Questions 1 through 3 refer to this information. Two types of basketball instruction were used with random samples of participants who either had previous experience playing or did not have previous experience. The means indicate proficiency at playing basketball at the end of treatment.

Type of Instruction			
	New	Conventional	Row Mean
Previous experience	$M = 230$	$M = 200$	$M = 215$
No previous experience	$M = 200$	$M = 230$	$M = 215$
Column Mean	$M = 215$	$M = 215$	

1. Does there seem to be a main effect for type of instruction?
2. Does there seem to be a main effect for experience?
3. Does there seem to be an interaction?

   Questions 4 through 6 refer to this information. Random samples of participants with back pain and headache pain were randomly assigned to two types of pain relievers. The means below indicate the average amount of pain relief. A higher mean indicates greater pain relief.

Type of Pain			
	Back Pain	Headache	Row Mean
Type A pain reliever	$M = 25.00$	$M = 20.00$	$M = 22.50$
Type B pain reliever	$M = 15.00$	$M = 10.00$	$M = 12.50$
Column Mean	$M = 20.00$	$M = 15.00$	

4. Does there seem to be a main effect for type of pain (back pain versus headache pain)?
5. Does there seem to be a main effect for type of pain reliever?
6. Does there seem to be an interaction?

   Questions 7 through 9 refer to this ANOVA table:

Source	F	p
Age level (young, old)	13.25	0.029
Region (north, south)	1.69	0.321
Interaction (age × region)	15.32	0.043

$^*p < 0.05$

7. Is the main effect for age level statistically significant at the 0.05 level?
8. Can the null hypothesis for the main effect of region be rejected at the 0.05 level?
9. Is the interaction between age and region statistically significant at the 0.05 level?

Question for Discussion

10. Briefly describe a hypothetical study in which it would be appropriate to conduct a Two-Way ANOVA.

Notes

1. Note that income in large populations is usually skewed, making the mean an inappropriate average (see Chapter 12); for these groups, assume that it was not skewed. Also note that the Row Means and Column Means were obtained by adding and dividing by 2; this is appropriate only if the number of participants in all cells is equal. If it is not, compute the row and Column Means using the original raw scores.
2. The basis for this second statement is that if you subtract across the rows, you get the same difference for each row. Earlier, you were instructed to subtract down columns; however, subtracting across the rows works equally well in determining whether there is an interaction.
3. It is also common to include the degrees of freedom, the sums of squares, and mean squares in an ANOVA table. As mentioned in Chapter 29, however, these are of little interest to the typical consumer of research.

PREDICTIVE SIGNIFICANCE

While there are various types of inferential statistics, we have discussed null hypothesis and significance tests around means comparison and One-Way and Two-Way ANOVA, which are parametric statistics. In this final part of the book, non-parametric statistics called Chi-Square (One-Way and Two-Way) is introduced. Chi-Square is used to measure associations when violation of assumption of normality of data is in place and data are categorical. We will discuss measuring the effect size of various statistics to deepen our understanding of significance testing. And finally, simple and multiple linear regression will be discussed around the predictability of a dependent variable by independent variables. The sequence of the topics in Part G is as follows:

Chapter 31. Chi-Square Test
Chapter 32. Effect Size
Chapter 33. Simple and Multiple Linear Regression

PREDICTIVE SIGNIFICANCE

Chi-Square Test

Chapter Objectives

The reader will be able to:

- ❏ Describe the use of One-Way and Two-Way Chi-Square Tests as a non-parametric statistic.
- ❏ Demonstrate how to report and interpret Chi-Square Test results.

All of the inferential statistics, including the topics discussed in this book, are based on a set of assumptions. One of these assumptions is that the data is normally distributed. But of course, not all data are normally distributed. Populations are sometimes skewed in their distribution. For example, the tests on means (t and F) in earlier chapters of this book are based on the assumption that the underlying distributions are normal; they are examples of ***parametric tests***. The good news is that even when the assumption of normality of data and other assumptions are violated, there are other sets of statistics that can be performed. These statistics are called ***non-parametric statistics***. Some of these non-parametric statistics are somewhat equivalent to parametric statistics, like Independent

Tests that are based on the assumption that the underlying distributions are normal are examples of ***parametric tests***.

When the assumption of normality of data and other assumptions are violated, there are other sets of statistics that can be performed. These statistics are called ***non-parametric statistics***.

DOI: 10.4324/9781003299356-38

Samples *t* Test, the Dependent Samples *t* Test, or One-Way ANOVA discussed earlier.

The most commonly used non-parametric statistics is the ***Chi-Square Test of Independence***. Since Chi-Square is not based on such an assumption, it is an example of a *non-parametric* (or *distribution-free*) test. The Chi-Square Test is used when the data are from two categorical or nominal variables. (See Chapter 3 for descriptions of types of variables.)

Frequently, research data are *nominal* (i.e., naming data, such as participants naming the political candidates for whom they plan to vote).[1] Because such data do not consist of scores, they do not directly permit the computation of means and standard deviations. Instead of reporting means and standard deviations for such data, researchers typically report the number of participants who named each category (e.g., named each political candidate) and the corresponding percentages. Example 1 below illustrates such a report.

EXAMPLE 1

A random sample of 200 registered voters was drawn and asked which of two candidates for elected office they planned to vote for. The data below indicate that the majority plans to vote for Candidate Smith.

Candidate Smith	Candidate Doe
n = 110 (55.0%)	*n* = 90 (45.0%)

To test differences among frequencies, researchers use *Chi-Square*. Although the data in Example 1 suggest that Candidate Smith is preferred, keep in mind that only a random sample of 200 registered voters was surveyed. Therefore, it is possible, for instance, that the population of likely voters is evenly split, but that a difference of ten percentage points (55% versus 45%) was obtained because of the sampling errors associated with random sampling. For this possibility, the *null hypothesis* states that there is no true difference in the population—that is, the population of registered voters is evenly split.

In Example 1, the *t* test and the *F* test (i.e., ANOVA) cannot be used to test the null hypothesis because they are tests of differences among means, but the results in the example do not consist of means. Instead, they consist of the numbers of cases (*n*) and percentages.[2]

The appropriate test for the data under consideration in Example 1 is *Chi-square*, whose symbol is $\chi 2$. A Chi-Square Test is designed to test for differences among frequencies. As it turns out, a Chi-Square Test for the data in Example 1 indicates that the probability that the null hypothesis is a correct hypothesis is greater than 5 in 100 ($p > 0.05$). Thus, the null hypothesis cannot be rejected and the difference between the frequencies (110 versus 90) cannot be declared statistically significant.

Note that while Chi-Square compares frequencies, whatever it determines about the statistical significance of the frequencies is also true of the associated percentages. Thus, because the difference between the two frequencies is not statistically significant, the difference between the percentages is also not statistically significant.

Example 1 contains data that call for a **One-Way Chi-Square Test** (also known as a *goodness of fit Chi-Square Test*). This is because the participants are classified in only one way: according to the candidate for whom they plan to vote.

In contrast, Example 2 contains data that call for a **Two-Way Chi-Square Test**. In this example, each participant is classified according to two ways: (1) whether the respondent is male or female and (2) the candidate for who they plan to vote.

In a **One-Way Chi-Square Test**, participants are classified in only one way.

In a **Two-Way Chi-Square**, participants are classified in two ways.

EXAMPLE 2

A random sample of 200 male voters and a random sample of 200 female voters were drawn and asked to name the candidate for whom they planned to vote. These data were obtained:

	Candidate Jones	Candidate Rodriguez
Male voters	$n = 80$	$n = 120$
Female voters	$n = 120$	$n = 80$

Inspection of the data in Example 2 suggests that Candidate Jones is a stronger candidate among female voters, while Candidate Rodriguez is a stronger candidate among male voters. This suggests that there is a relationship between male/female categories and preference for candidates. If this pattern is true among all male and female voters in the population, both candidates should take heed. For instance, Candidate Jones might consider ways to shore up her support among male voters without alien-ating the female population, while Candidate Rodriguez might do the opposite.

However, only a random sample was surveyed in Example 2. Thus, before acting on the results of the survey, the candidates should consider how likely it is that the observed differences in preferences between the two groups (males and females) were created by random sampling errors (i.e., the assertion made by the null hypothesis). As it turns out, for Example 2, the Two-Way Chi-Square Test reveals that the probability that the null hypothesis is correct is less than 1 in 1,000 ($p < 0.001$). Thus, it is very unlikely that this pattern of differences is due to sampling errors. Thus, with a high degree of confidence, the candidates can rule out sam-pling error as an explanation for the pattern of differences between male and female voters.[3]

In Example 3, a sample from one population of participants was asked two questions, each of which provided nominal data.

EXAMPLE 3

A random sample of college students was asked (1) whether they think that IQ tests measure innate (i.e., inborn) intelligence and (2) whether they had taken a course in psychological testing. These data showed the following:

	Took Course	Did Not Take Course
Yes, innate intelligence	$n = 20$	$n = 30$
No, not innate intelligence	$n = 40$	$n = 15$

The data in Example 3 suggest a relationship between whether students have taken the course and what they believe IQ tests measure.

The data in Example 3 suggest that those who took the course are less likely to think that IQ tests measure innate intelligence (20 "yes" versus 40 "no") than those who did not take the course (30 "yes" versus 15 "no"). Thus, there appears to be a relationship between whether participants have taken the course and whether they believe IQ tests measure innate intelligence. However, only a random sample was questioned and, thus, it is possible that the observed relationship is only due to sampling error. That is, the null hypothesis asserts that there is no *true* relationship (in the population). A Chi-Square Test for the data in Example 3 produced this result:

$$\chi^2 = 11.455, \, df = 1, \, p < 0.001$$

Because $p < 0.001$, the null hypothesis can be rejected with a high degree of confidence. Note that the odds are less than 1 in 1,000 that the null hypothesis for the data in Example 3 is a correct hypothesis.[4]

Example 4 shows how the results of the Chi-Square Test for Example 3 might be stated in a research report in an academic journal.

EXAMPLE 4

The relationship was statistically significant with those who took a course in psychological testing being less likely to believe that IQ tests measure innate intelligence than those who did not take the course ($\chi^2 = 11.455$, $df = 1$, $p < 0.001$).

The Chi-Square Test for Example 3 indicates that the null hypothesis should be rejected at the 0.001 level; the relationship is statistically significant.

Exercise for Chapter 31

Factual Questions

1. If you calculated the mean math test score for freshmen and the mean math test score for seniors and wanted to compare the two means for statistical significance, would a Chi-Square Test be appropriate? Explain.
2. If you asked members of a random sample which of two types of skin cream they prefer, and you wanted to compare the resulting frequencies with an inferential statistical test, would a Chi-Square Test be appropriate?
3. If you asked members of a random sample (1) which of two types of skin cream they prefer and (2) whether they were satisfied with the condition of their skin, would a "One-Way Chi-Square Test" *or* a "Two-Way, Chi-Square Test" be appropriate?
4. For examining relationships for nominal data, should a researcher use a "One-Way Chi-Square Test" *or* a "Two-Way Chi-Square Test"?
5. Suppose you read that "$\chi^2 = 4.111$, $df = 1$, $p < 0.05$." What decision should be made about the null hypothesis at the 0.05 level?
6. Suppose you read that "$\chi^2 = 7.418$, $df = 1$, $p < 0.01$." Is this statistically significant at the 0.01 level?
7. Suppose you read that "$\chi^2 = 2.824$, $df = 2$, $p > 0.05$." What decision should be made about the null hypothesis at the 0.05 level?

8. If as a result of a Chi-Square Test, p is found to be less than 0.001, the odds that the null hypothesis is correct are less than 1 in _____?

Question for Discussion

9. If you asked members of a random sample whether they planned to vote "yes" or "no" on a ballot proposition, would a "One-Way Chi-Square Test" *or* a "Two-Way Chi-Square Test" be appropriate? If you answered, One-Way Chi-square, how would you build the study into a Two-Way Chi-Square Test?

Notes

1. See Chapter 3 to review the meaning of the nominal scale of measurement.
2. The tests on means (t and F) in earlier chapters of this book are based on the assumption that the underlying distributions are normal; they are examples of *parametric tests*. Since Chi-Square is not based on such an assumption, it is an example of a *nonparametric* (or *distribution-free*) test.
3. Example 2 illustrates that a Two-Way Chi-Square is used to test for a *relationship* between two variables. In this example, there is a significant relationship between voters' preferences for a candidate and whether they are male or female voters.
4. Notice that in the examples, the responses are independent. For instance, in Example 2, the sex of a person is not determined by his or her preference for a candidate. Also, each response category is mutually exclusive. For instance, a participant is not allowed to indicate that he/she is both sexes. Independent and mutually exclusive categories are assumptions underlying the Chi-Square Test.

CHAPTER 32

Effect Size

Chapter Objectives

The reader will be able to:

❏ Describe the effect size of a statistically significant test using *Cohen's d*.
❏ Explain that effect size standardizes the significant difference between two means.
❏ Explain that effect size is interpreted relative to the standard deviation units of the control group.
❏ Recognize the labels for values of *d* and the principles and interpretation of effect size.

The first of two common tools used to reach conclusions is how well statistical results from a sample generalize to the large population and is statistical significance. The second of these is ***effect size***, which, like statistical significance helps us to understand how meaningful the results are. Suppose a researcher uses a test or scale with which you are very familiar. When the researcher reports that the difference between the experimental and control groups is such-and-such a number of points, you will have a good idea of how large and important the difference is.

Effect size
standardizes the size of the difference between two means.

DOI: 10.4324/9781003299356-39

In ANOVA, effect size is measured by η^2.

However, very often you will not be familiar with the tests and instruments used in research. This makes it difficult to assess whether the difference is "large." Thus, *effect size* measures how large an observed effect is by standardizing the size of the difference between two means.

Effect size is measured differently with different statistics. For example, in correlational study or regression analysis, it is common to use the coefficient of determination (r^2 or R^2) as discussed in Chapter 19. In ANOVA, when we compare two or more means, the effect size is measured by η^2 as discussed in Chapter 29. In this chapter, we will discuss a specific measure called **Cohen's d** that is used to measure the effect size for *t* test of means comparisons. See Table 32.1 for a summary of various measures of effect size.

TABLE 32.1
A Summary Table of Measures of Effect Size by Statistical Procedure

Statistical Procedure	Effect Size	What Is Measured	Calculation
Correlational study (to measure the relationship between two variables)	r^2	Measures how much variance in one variable (dependent or outcome variable) is explained by one variable (independent or predictor variable)	Coefficient of Determination = Correlation coefficient r squared (See Chapter 19)
Regression analysis (to measure the predictive relationship among variables)	R^2	Measures how much variance in one variable (dependent or outcome variable) is explained by a group of variables (independent or predictor variables)	Coefficient of Determination = Multiple correlation coefficient R squared Range: 0 to 1.00 (0%–100%) (See Chapter 20)
ANOVA (to compare two or more means)	η^2	Measures how much of a variable is explained by group differences	$\dfrac{\text{Between Groups Sums of Squares}}{\text{Total Sums of Squares}}$ Range: 0 to 1.00 (0%–100%) (See Example in Table 29.1)
t test (to compare two means)	Cohen's d	Standardizes the difference between two means	$d = \dfrac{Mean_1 - Mean_2}{\sqrt{pooled\ sample\ variance}}$ Range: $0 - 4$ s.d. units

Cohen's *d* (also labeled as *d*) is one of the most widely used statistics for the purpose of standardizing the difference between two means to measure the effect size, which is quite easy to compute. Simply subtract the control group mean from the experimental group mean and divide by the square root of the pooled sample variance, where the variances (sd²) of two groups are averaged and square rooted.[1] The following is an example. Suppose Researcher A used a scale with possible score values from 0 to 100 and obtained these results:

Experimental Group: M = 40
Control Group: M = 30,
where the *square root of pooled sample variance = 10*
Cohen's d = (40 – 30)/10 = 1

If we read a report that states that *d* equals 1.00 (based on the statistics shown immediately above), what does this indicate? Simply this: the mean of the experimental group is a full standard deviation higher than the mean of the control group. Is this large? Yes, it is quite large. Remember that there are only about three standard deviation units above (and below) the mean and that the vast majority of cases are within one standard deviation of the mean. Hence, if the average (mean) participant in the experimental group is a full standard deviation above the average participant in the control group (i.e., *d* = 1.00), the average experimental group participant is higher than the vast majority of those in the control group.

The term *standardizing* was used earlier. As it turns out, using *d* standardizes descriptions of the sizes of differences. To illustrate this, consider Researcher B, who studied the same phenomena as Researcher A in the earlier example, but used a scale with scores that could range from 200 to 800 and obtained these statistics:

Experimental Group: M = 400
Control Group: M = 300,
where the *square root of pooled sample variance = 100*
Cohen's d = (400 – 300)/100 = 1

Notice that the value of *d* is the same in both examples even though very different scales were used. Researcher A got a 10-point raw difference while Researcher B got a 100-point raw difference.

However, by dividing the corresponding control group's standard deviation into each of these differences, we find that the *effect size* (standardized size of the difference) is the same (*d* = 1.00) in both cases. This

Cohen's d is a widely used measure of effect size for *t* test of means comparisons.

There are only about three standard deviation units above (and below) the mean.

Cohen's d expresses each difference in terms of standard deviation units.

standardization occurs because *d* expresses each difference in terms of standard deviation units, which usually range from three below to three above the mean. Put another way, using *standard* deviation units as a way of looking at differences *standardizes* the process.

While there are no universally accepted standards for describing values of *d* in words, many researchers that use Cohen's (1992)[2] suggests the following:

(1) *Effect size* of *d* of about 0.20 (one-fifth of a standard deviation) is considered "small."
(2) *Effect size* of 0.50 (one-half of a standard deviation) is considered "medium."
(3) *Effect size* of 0.80 (eight-tenths of a standard deviation) is considered "large."

(Keep in mind that in terms of values of *d*, an experimental group can rarely exceed a control group by more than 3.00 because the effective range of standard deviation units is *mainly* only 3 on each side of the mean. Thus, for most practical purposes, 3.00 [or −3.00] is the maximum value of *d*.)[3]

Extrapolating from Cohen's suggestions, a value of 1.10 might be called "very large" and a value of 1.40 or more might be called "extremely large." Values this large are rarely found in social and behavioral research.

Effect size of 1.10 is considered very large.
Effect size of 1.40 or greater is considered extremely large.

Table 32.2 summarizes these guidelines for interpreting values of *d*.

The labels being discussed should not be used arbitrarily without consideration of the full context in which the values of *d* were obtained and the possible implications of the results. This leads to two principles: (1) a small effect size might represent an important result, and (2) a large effect size might represent an unimportant result.

In terms of values of *d*, an experimental group rarely exceeds a control group by more than 3.00.

TABLE 32.2
Labels for Values of *d*

Value of *d*	Label
0.20	Small
0.50	Medium
0.80	Large
1.10	Very large
1.40+	Extremely large

Consider the *first principle*: a small effect size might represent an important result. Suppose, for instance, that researchers have been frustrated by consistently finding values of *d* well below 0.20 when trying various treatments for solving an important problem (such as treatments for a new and deadly disease). If a subsequent researcher finds a treatment that results in a value of about 0.20, this might be considered a very important finding. At this low level (0.20), the effect of the treatment is small but it might be of immense importance to sick individuals helped by the treatment—however small the effect size is. In addition, the results might point the scientific community in a fruitful direction with regard to additional research on treatments for the problem in question.

The *second principle* is that a large value of *d*—even one above 1.40—might be of limited importance. This is most likely when the results lack practical significance in terms of cost, public and political acceptability, and ethical and legal concerns. (See Chapter 23 for considerations in determining the practical significance of research results.)

Here are three steps for interpreting the difference between two means.

Step 1. Determine whether the difference is statistically significant at an acceptable probability level, such as $p < 0.05$. If it is not, the difference should usually be regarded as unreliable and should be interpreted as such.

Step 2. If there is a statistically significant difference, consider the value of *d* and consider the labels in Table 32.2 for describing the magnitude of the difference.

Step 3. Consider the implications of the difference for validating any relevant theories as well as the practical significance of the results.

Of course, before researchers begin to follow the three steps outlined above, the adequacy of the research methodology that they employed should be considered. Woefully inadequate sampling (such as a very biased sample), clearly invalid instrumentation (such as a test that measures a variable other than the one the researcher wanted to study), or a very poor research design (such as a design that will not answer the research question) would lead to very serious questions regarding the validity of the results. In such cases, consideration of values of *d* might be meaningless. (See Appendix A.)

The first principle of *effect size*: a *small* effect size might represent an important result.

The second principle of *effect size*: a *large* effect size might represent an unimportant result.

Consider the value of *d* for a statistically significant difference.

Consider the adequacy of the research methodology when considering *effect size*.

Exercise for Chapter 32

Factual Questions

1. For an experimental group, $M = 50$. and for the control group, $M = 46$, with the *square root of pooled sample variance of 8*. For this experiment, what is the value of d?

2. Is the effect size for Question 1 "very large"?

3. If the value of d for the difference between two means equals 1.00, the experimental group's mean is how many standard deviation units higher than the control group's mean?

4. What value of d is associated with the label "extremely large"?

5. According to Cohen (1992), what label should be attached to a value of d of 0.80?

6. Under what circumstance will a negative value of d be obtained?

7. Should a test of statistical significance be conducted "before" *or* "after" d is computed and its value interpreted with labels?

Question for Discussion

8. As noted in this topic, a small value of d might be associated with an important result. Name a specific research result scenario where even a small value of d might indicate a result of great practical importance.

Notes

1. See Appendix A for the formulas for Cohen's *d for one sample using the SD of population and two samples* using the pooled sample variance as the divisor.
2. Cohen, J. (1992). A power primer. *Psychological Bulletin, 112*, 155–159.
3. A negative is obtained when the control group's mean is higher than the experimental group's mean. Note that less than one-half of 1 percent of a normal distribution lies above +3.00 and below −3.00, which means that it is technically possible—but highly unlikely—to obtain values larger than +3.00 and −3.00.

CHAPTER 33

Simple and Multiple Linear Regression

Chapter Objectives

The reader will be able to:

- ❏ Describe the concept of simple linear regression as a way to predict one variable on the basis of another.
- ❏ Describe the concept of multiple regression as a way to predict one variable on the basis of multiple variables.
- ❏ Identify how results of regression analysis are reported and interpreted.

The concept of correlation was discussed in Part D as a measure of how two or multiple variables are related to each other. This led to a discussion about a closely related concept of coefficient of determination, which measures how much of the relationship between two variables can be explained in terms of the percentage of variance in one variable explaining the variance in the other variable (review Chapter 19). The same concept of correlation and coefficient of determination applies to multiple correlations that involve a relationship between a combination of variables to increase predictability of the predicted variable.

DOI: 10.4324/9781003299356-40

Simple linear regression and *multiple linear regression* are very common tools used to analyze the predictive power of one or more independent variables on a dependent variable.

The higher the correlation coefficient, the more predictive it is between one variable and the other.

We allocate the independent variable that usually occurs first as "predictor(s)" to be represented by X and the dependent variable or the "predicted" or "outcome" to be represented by Y.

One independent variable predicting one dependent variable is known as *simple linear regression*.

Most often when we find a significant correlation between variables or among several variables, we want to establish predictions to generalize for future populations. Recall that correlation measures relationship without distinguishing which variable is independent or dependent because correlation measures the degree of association between variables without predictability or causality between them. But by using statistical tools called *simple linear regression* and *multiple linear regression*, which are very common tools in the social sciences, we can analyze the predictive power of one or more independent variables on a dependent variable. And in the case of multiple independent variables predicting the dependent variable, we can understand the predictability of these variables together and individually in predicting the dependent variable. In other words, we use correlations and scattergrams to measure how much of the relationship is predictive or causal. The higher the correlation coefficient, the more predictive it is between one variable and the other.

For example, based on a significant correlation between students' math test scores and their math aptitude test scores, let's say that we want to know whether the former predicts the latter. For instance, a student named Kirsten scores 25 on her recent math test, and John scores 75 on his recent math test. We can predict that Kirsten's math aptitude test score will be low and John's test score will be high. The basic idea is to use a set of data on two variables, X and Y measuring how much X predicts Y. And if the correlation between the X and Y variables is high, the stronger the predictability is.

An independent variable is labeled, X, as a placeholder for any variable of your interest to predict the dependent variable, Y. These symbols are easy to remember. Just like the letter X comes first, then the letter Y in the alphabet, we allocate the independent variable that usually occurs first as the "predictor(s)" to be represented by X and the dependent variable or the "predicted," or "outcome," to be represented by Y.

In the above example, the students' math test scores would be the independent variable, X, and their math aptitude test scores would be the dependent variable, Y.

In another example, let's say we establish a strong correlation between college freshmen SAT scores and their first-year college GPA at a local state college. Then through a simple linear regression analysis, the researcher sees that SAT scores predict college GPA with high accuracy. Based on this analysis, the researcher should be able to predict what first-year college GPA would be by their SAT scores for all students being admitted in the future.

All of these examples so far have been for one independent variable predicting one dependent variable. This is known as *simple linear regression*.

But does having multiple independent variables predicting the dependent variable have more predictive power? Having multiple indicators would give us a more comprehensive understanding of the dependent variable.

For example, suppose a high school algebra teacher wants to understand what explains high achievement among her math students. High achievement would be the dependent variable. She observed that a positive experience with math in the previous year was highly correlated with high achievement in her classroom. But explaining that high achievement can be complex, as multiple attributing factors may explain high achievement, such as the students' prior exposure to math, their attitudes toward math, and/or home environment. All these and numerous other factors may influence students' achievement in the math class. If we see that the measured indicators show high correlation to the high achievement, then we can see how all of the measured independent variables together correlate with the dependent variable in a multiple correlation (see Chapter 20). For predictability, since there is more than one independent variable predicting the dependent variable, we would use ***multiple linear regression***.

A word of caution. We want to make sure that these possibly significant independent variables would contribute to explaining the dependent variable before including them in the analysis. For example, if in the above example, students' prior exposure to math is highly correlated with their attitude toward math, you would have two independent variables that may be very similar. Then you may not need to include both in the multiple regression.

Here are some general rules about selecting an independent variable(s).

1. When selecting an independent variable in a simple linear regression, make sure that the variable is conceptually related to the dependent variable and shows a significant correlation.
2. When selecting multiple independent variables in a multiple linear regression, you want to make sure that the independent variables are least correlated with each other, but individually and together are highly correlated to the dependent variable.
3. Independent variables can be either categorical or continuous data, predicting a dependent variable that is continuous data.[1]
4. Generally, the more comprehensive you could be by adding more meaningful independent variables, the better (but oftentimes, time and resources limit data collection and analysis).

If there is more than one independent variable predicting the dependent variable, we would use ***multiple linear regression***.

Since there is more than one independent variable predicting one dependent variable, we would use *multiple regression*.

How do we show results and interpret them? Let's look at another example. A researcher interested in school climate and students' achievement conducted a study. The research question was: What type of school climate is most conducive to student achievement? School climate was measured by multiple measures that were summarized into six areas or constructs listed below, and students' achievement was measured by their GPA.[2] They were:

1. *Physical Appearance of the School* (PA), such items as: things get fixed immediately.
2. *Student Interactions* (SI), such as: students feel like they are part of the school family.
3. *Discipline Environment* (DE), such as: classrooms are a positive place.
4. *Leadership/Assessment* (LA), such as: instruction is dynamic, involving, learner-centered, and challenging.
5. *Attitude and Culture* (AC), such as: students feel a sense of belonging, school maintains traditions that promote school pride.
6. *School–Community Relations* (SR), such as: school is welcoming of others in the community.

Table 33.1 shows a correlation matrix among the six independent variables and GPA using partial hypothetical data.

The last column in the matrix shows that all independent variables, except PA (physical appearance), are significantly correlated to GPA, as noted by "*" or "**" by the correlation coefficients. But you also see that the independent variables are significantly related to each other as noted by "**" in other columns.

The results showed a multiple correlation R of 0.721, and R^2 of 0.520, $F_{(6.31)} = 5.591$, $p < 0.01$. This indicates that 52% of the variance in GPA is explained by the six independent variables overall. Table 33.2 reports the regression coefficients for each independent variable. The results show that only one variable, SR, significantly predicted GPA. Its probability value (p) is less than 0.001.

Like in other inferential statistics discussed thus far, the probability value (p) for each predictor variable shows significance. While all but one predictor was significantly correlated with GPA, the regression analysis revealed that only SR significantly predicted GPA. Notice that the t value associated with SR is the highest among all the t values in the analysis.[3] While the results of this example are from fictitious data, the tables illustrate how the results of a multiple regression analysis are reported and interpreted.

TABLE 33.1

Correlations of the Variables

	PA	SI	DE	LA	AC	SR	GPA
PA	–	0.713**	0.512**	0.522**	0.492**	0.409**	0.262
SI		–	0.607**	0.452**	0.664**	0.629**	0.301*
DE			–	0.564**	0.611**	0.656**	0.465**
LA				–	0.630**	0.583**	0.406**
AC					–	0.759**	0.432**
SR						–	0.718**
GPA							–

* $p < .05$, ** $p < .01$

TABLE 33.2

Results of Multiple Regression Analysis

Variables	B	β	t	p
PA	0.083	0.091	0.464	0.646
SI	0.082	0.101	0.521	0.606
DE	−0.031	−0.041	−0.231	0.819
LA	0.040	0.057	0.313	0.756
AC	−0.227	−0.289	−1.370	0.181
SR	0.657	0.793	4.203	0.000***

*** $p < 0.001$

Exercise for Chapter 33

Factual Questions

1. Discuss how a regression analysis can develop from correlation coefficients.
2. What is the main distinction between correlation and regression in terms of their purpose?
3. Discuss the difference between simple linear regression and multiple regression in terms of the types of variables for each analysis.
4. If two independent variables show a high correlation in a multiple regression, what can you suspect in terms of their predictability of the dependent variable?

Question for Discussion

5. Discuss a possible study topic that would require a simple linear regression, then develop it into a multiple regression.

Notes

1. If the dependent variable is categorical data with two categories called dichotomous variable, logistic regression analysis is performed, and with two or more categories, discriminant analysis. This is beyond the scope of this book.
2. The instrument is adapted from Shindler et al. (2016). School climate–student achievement connection: if we want achievement gains, we need to begin by improving climate. *Journal of School Administration Research and Development, 1(1),* 9–16.
3. The B and ß coefficients give additional information about the individual predictor variables. The B coefficient is a slope, a rate of increase in the dependent variable per every increase in the independent variable. And because B coefficients are on different scales, the ß coefficients show standardized Bs. Therefore, comparing the ß coefficients shows rank among variables in the order of strength of predictability, regardless of negative/positive signs.

Appendix A
Computations

Chapter 8 Computation of the Standard Error of the Mean

Formula:

$$SE_M \frac{s}{\sqrt{N}} \qquad\qquad \text{a1}$$

Chapter 14 Computation of the Standard Deviation

To compute the standard deviation for a large number of scores, a computer should be used. However, Considering how the standard deviation is computed, using a formula can help in understanding its meaning.

The formula that defines the standard deviation is:

$$s = \sqrt{\frac{\Sigma(X-M)^2}{N}} \qquad\qquad \text{a2}$$

This formula is also known as the definitional formula.

The formula *(x – M)* stands for the deviation of each score from the mean of its distribution. To obtain it, first calculate the mean (in this case, the sum of all Xs is 78 and $N = 6$, thus, $78/6 = 13.00$) and subtract the mean from each score, as shown in Example 1. Deviations show how far each score is from the mean. Then square the deviations and sum the squares, as indicated by the symbol Σ. Then enter this value in the formula along with the number of cases *(N)* and perform the calculations as indicated here.

Here's an example of calculating the standard deviation using the definitional formula.

EXAMPLE 1

Scores (X)	Deviations ($\bar{X}$ – M)	Squared Deviations ($\bar{X}$ – M)²
10	10 – 13.00 = –3	9.00
11	11 – 13.00 = –2	4.00
11	11 – 13.00 = –2	4.00
13	13 – 13.00 = 0	0.00
14	14 – 13.00 = 1	1.00
19	19 – 13.00 = 6	36.00
		$\Sigma(\bar{X} - M)^2 = 54.00$

Thus, substituting the sum of the squared deviations (54) and the number of cases (6 scores) in the formula:

$$s = \sqrt{\frac{\Sigma(X-M)^2}{N}} = \frac{54}{6} = \sqrt{9.00} = 3.00 \qquad \text{a3}$$

As the formula indicates, the standard deviation is the *square root of the average squared deviation from the mean.* Thus, the larger the deviations from the mean, the larger the standard deviation. Conversely, the smaller the deviations from the mean, the smaller the standard deviation. At the extreme, when all the scores are the same, the standard deviation equals zero, as indicated in Example 2, where each score is 20 and therefore the mean of the scores is also 20 (i.e., 20.00).

EXAMPLE 2

Scores (X)	Deviations $(X - M)$	Squared Deviations $(\bar{X} - M)^2$
20	$20 - 20.00 = 0$	0.00
20	$20 - 20.00 = 0$	0.00
20	$20 - 20.00 = 0$	0.00
20	$20 - 20.00 = 0$	0.00
20	$20 - 20.00 = 0$	0.00
20	$20 - 20.00 = 0$	0.00
		$\Sigma(\bar{X} - M)^2 = 0.00$

Thus, substituting the sum of the squared deviations (0.00) and the number of cases (6 scores) in the formula:

$$S = \sqrt{\frac{0.00}{6}} = \sqrt{0.00} = 0.00 \qquad\qquad \text{a4}$$

Thus, when there is no variation (i.e., all scores are the same), the standard deviation equals 0.00.

The formula discussed is known as the definitional formula that illustrates the concept of standard deviation within the formula. For computational purposes, there is an easier formula known as the *computational formula of standard deviation*.

Computational formula for standard deviation of a sample:

$$S = \sqrt{\frac{\Sigma(X^2) - (\Sigma X)^2 / N}{N}} \qquad\qquad \text{a5}$$

For example:

Scores (X)	X^2
10	100
11	121
11	121
13	169
14	196
19	361
$\Sigma x = 78.00$	$\Sigma(x^2) = 1,068$

Here's the calculation.

$$S = \sqrt{\dfrac{1068 - (78)^2 / 6}{6}}$$

$$= \sqrt{\dfrac{1068 - (6084)/6}{6}}$$

$$= \sqrt{\dfrac{1068 - 1014}{6}} = \sqrt{\dfrac{54}{6}} = \sqrt{9.00} = 3.00$$

Chapter 17 Computation of Pearson *r*

Formula:

$$r = \dfrac{\left[N(\Sigma XY) - (\Sigma X)(\Sigma Y) \right]}{\sqrt{\left[N(\Sigma X^2) - (\Sigma X)^2 \right]\left[N(\Sigma Y^2) - (\Sigma Y)^2 \right]}}$$

Scores (X)	Scores (Y)	(XY)	X^2	Y^2
3	6	18	9	36
5	8	40	25	64
4	9	36	16	81
9	10	90	81	100
10	12	120	100	144
$\Sigma X = 31$	$\Sigma Y = 45$	$\Sigma XY = 304$	$\Sigma X^2 = 231$	$\Sigma Y^2 = 425$

From the table above, we have

ΣX	=	31
ΣY	=	45
ΣXY	=	304
ΣX^2	=	231
ΣY^2	=	425

Calculation:

$$r = \dfrac{\left[N(\Sigma XY) - (\Sigma X)(\Sigma Y) \right]}{\sqrt{\left[N(\Sigma X^2) - (\Sigma X)^2 \right]\left[N(\Sigma Y^2) - (\Sigma Y)^2 \right]}}$$

$$r = \dfrac{\left[5(304) - (31)(45) \right]}{\sqrt{\left[5(231) - (31)^2 \right]\left[5(425) - (45)^2 \right]}}$$

$$r = \frac{[1520 - 1395]}{\sqrt{1155 - 961][2125 - 2025]}}$$

$$r = \frac{125}{\sqrt{[194][100]}} = \frac{125}{\sqrt{[19400]}} = \frac{125}{139.28} = .90 \qquad \text{a8}$$

Chapter 25 Computation of Independent Samples *t* Test

$$t = \frac{\text{Difference between two independent sample means}}{\text{Standard error of the difference between means}} \qquad \text{a9}$$

where standard error of the difference between means is,

$$S_D = \sqrt{\left[\frac{(n_1 - 1)s_1^2 + (n_2 - 1)s_2^2}{n_1 + n_2 - 2}\right]\left[\frac{n_1 + n_2}{n_1 n_2}\right]} \qquad \text{a10}$$

and degrees of freedom $(df) = n_1 + n_2 - 2$

A Summary Table of Computation Formula

Means Comparison *t* Tests	Computation Formula
Independent Samples *t* Test	$t = \dfrac{\text{Difference between two independent sample means}}{\text{Standard error of the difference between means}}$
Dependent Samples *t* Test	$t = \dfrac{\text{Difference between two related sample means}}{\text{Standard error of the difference between means}}$
One-Sample *t* Test	$t = \dfrac{\text{Sample mean} - \text{Population mean}}{\text{Standard error of means}}$
	where *Standard Error of Mean* $= \dfrac{sd}{\sqrt{n}}$

Chapter 26 Computation of Dependent Samples *t* Test

$$t = \frac{\text{Difference between two dependent sample means}}{\text{Standard error of the difference between means}} \qquad \text{a11}$$

where standard error of the difference between means is,

$$S_{\bar{D}} = \sqrt{\frac{n\sum(D^2) - \sum(D^2)}{n-1}} \qquad \text{a12}$$

and degrees of freedom (*df*) = *N* – 1
   (*D* indicates mean difference)

Chapter 27 Computation of One-Sample *t* Test

$$t = \frac{\bar{x} - \mu}{SE_M}, \text{ where } SE_M = \frac{s}{\sqrt{n}}$$

and degrees of freedom (*df*) = *N* – 1

$$\frac{20-22}{4/\sqrt{40}} = \frac{-2}{4/6.32} = \frac{-2}{.633} = -3.16$$

Chapter 32 Effect Size

Calculation of Cohen's *d* for one sample:

$$d = \frac{\text{Sample Mean} - \text{Population Mean}}{\text{Standard Deviation}}$$

Calculation of Cohen's *d* for two samples with equal sample sizes:

$$d = \frac{\text{Mean}_1 - \text{Mean}_2}{\sqrt{\text{pooled sample variance}}}$$

Where $\sqrt{\text{pooled sample variance}} = \sqrt{\frac{S_1^2 - S_2^2}{2}}$

Thus,

$$d = \frac{Mean_1 - Mean_2}{\sqrt{\dfrac{S_1^2 - S_2^2}{2}}}$$

Note that Cohen's d for two samples with unequal sample sizes need to weigh each sample variance by its sample size.

Appendix B
Notes on Interpreting Pearson *r* and Linear Regression

This appendix describes why a Pearson *r* may be misleadingly low and introduces a statistical procedure for making predictions for individuals when there is a reasonably strong relationship as indicated by a Pearson *r*.

Why the Pearson *r* May Be Misleadingly Low

The value of a Pearson *r* can be misleadingly low for two reasons. First, its value is diminished if the variability in a group is artificially low. For the sake of illustration, assume that we wanted to study the relationship between height and weight in the adult population but foolishly selected only participants who were all exactly six feet tall. When weighing them, we would undoubtedly find some variation in their weights. What is the correlation between height and weight among such a group? Even though there is a positive relationship between the two variables in the general adult population, the correlation in this odd sample is zero. This must be the result because those who weigh more and those who weigh less are all of the same height. (This sample cannot show that those who are taller tend to weigh more because all subjects are of the same height. Thus, the value of the Pearson *r* will equal 0.00.)

A more realistic example is the relationship between scores on a college admissions test and grades earned in college. Although the test is given to all applicants in order for admissions decisions about all of them to be made, grades are available only for those who were admitted, and the correlation between scores and grades can be computed only for those applicants on whom we have complete data. Unfortunately, those for whom we have complete data are those who tend to have higher scores. Thus, the scores of those who are admitted are less variable than the scores of all applicants (since the low-scoring applicants were not admitted). As a result, the value of the Pearson r will be lower than would be obtained if we correlated using scores and grades for *all* applicants.

The second reason that an r can be misleadingly low is if r is computed for a curvilinear relationship. For instance, the relationship between test-taking anxiety and performance on standardized tests might be curvilinear. That is, small amounts of anxiety might be beneficial in motivating examinees to do well on a test, but larger amounts of anxiety might be detrimental. Thus, as anxiety increases, up to a point there is a positive relationship with anxiety; after a critical point is reached, as anxiety increases there is a negative relationship. If the Pearson r is computed for such data, the negative part of the relationship will cancel out the positive part, yielding an r near zero. Pearson recognized this problem and warned against using his statistic for describing curvilinear relationships. Other techniques, such as the *correlation ratio*, which are beyond the scope of this book, are available for describing curvilinear relationships. Fortunately, such relationships are relatively rare in the social and behavioral sciences.

Making Specific Predictions for Individuals

Suppose you found a reasonably high value of the Pearson r, such as $r = 0.60$, between scores on a college admissions test and grades earned in a college.[1] This indicates that the admissions test is a reasonably valid predictor of grades. It does not indicate, however, how to make specific predictions for individuals who might apply to the college in the future. For example, if Marilyn has an admissions test score of 600, the Pearson r of 0.60 does not tell us what specific grade point average to predict for Marilyn. *Linear regression* is a statistical technique that enables us to make such predictions under most circumstances. It is beyond the scope of this book to describe this procedure, but students who have mastered correlational concepts may wish to pursue it in other books, including *Success at Statistics*, which is available from Routledge.

Note

1. In practice, values of the Pearson r for relationships between admissions test scores and grades earned in college rarely exceed 0.60 and are often substantially lower.

Appendix C
Table of Random Numbers

Row #																	
1	2	1	0	4	9	8	0	8	8	8	0	6	9	2	4	8	2	6
2	0	7	3	0	2	0	4	8	2	7	8	9	8	2	2	9	7	1
3	4	4	9	0	0	2	8	6	2	6	7	7	7	3	1	2	5	1
4	7	3	2	1	1	2	0	7	7	6	0	3	8	3	4	7	8	1
5	3	3	2	5	8	3	1	7	0	1	4	0	7	8	9	3	7	7
6	6	1	2	0	5	7	2	4	4	0	0	6	3	0	2	8	0	7
7	7	0	9	3	3	3	7	4	0	4	8	8	9	3	5	8	0	5
8	7	5	1	9	0	9	1	5	2	6	5	0	9	0	3	5	8	8
9	3	5	6	9	6	5	0	1	9	4	6	6	7	5	6	8	3	1
10	8	5	0	3	9	4	3	4	0	6	5	1	7	4	4	6	2	7
11	0	5	9	6	8	7	4	8	1	5	5	0	5	1	7	1	5	8
12	7	6	2	2	6	9	6	1	9	7	1	1	4	7	1	6	2	0
13	3	8	4	7	8	9	8	2	2	1	6	3	8	7	0	4	6	1
14	1	9	1	8	4	5	6	1	8	1	2	4	4	4	2	7	3	4
15	1	5	3	6	7	6	1	8	4	3	1	8	8	7	7	6	0	4
16	0	5	5	3	6	0	7	1	3	8	1	4	6	7	0	4	3	5

Row #																	
17	2	2	3	8	6	0	9	1	9	0	4	4	7	6	8	1	5	1
18	2	3	3	2	5	5	7	6	9	4	9	7	1	3	7	9	3	8
19	8	5	5	0	5	3	7	8	5	4	5	1	6	0	4	8	9	1
20	0	6	1	1	3	4	8	6	4	3	2	9	4	3	8	7	4	1
21	9	1	1	8	2	9	0	6	9	6	9	4	2	9	9	0	6	0
22	3	7	8	0	6	3	7	1	2	6	5	2	7	6	5	6	5	1
23	5	3	0	5	1	2	1	0	9	1	3	7	5	6	1	2	5	0
24	7	2	4	8	6	7	9	3	8	7	6	0	9	1	6	5	7	8
25	0	9	1	6	7	0	3	8	0	9	1	5	4	2	3	2	4	5
26	3	8	1	4	3	7	9	2	4	5	1	2	8	7	7	4	1	3

Comprehensive Review Questions

Chapter 1: The Empirical Approach to Knowledge

1. The *empirical* approach to knowledge is based on
   A. deduction.
   B. reliance on authority.
   C. observation.
2. "Everyday observation is an example of the empirical approach to knowledge." This statement is
   A. true.
   B. false.
3. If there are 800 teachers in a school district and 100 are selected for observation, the 100 are known as a
   A. population.
   B. sample.
4. "Flawed research can be as misleading as everyday observations." This statement is
   A. true.
   B. false.
5. What is a primary function of statistical analysis?
   A. Planning when observations will be made.

B. Organizing and summarizing data.

C. Identifying a population.

Chapter 2: Types of Empirical Research

1. Treatments are given in which type of study?
   A. Experimental.
   B. Non-experimental.
2. Treatments constitute which type of variable?
   A. Independent.
   B. Dependent.
3. Suppose students were treated with two types of rewards to see which one was more effective in promoting spelling achievement. *Spelling achievement* is the
   A. independent variable.
   B. dependent variable.
4. Researchers try to change the participants in which type of study?
   A. Experimental.
   B. Descriptive.
5. A survey is an example of
   A. an experimental study.
   B. a non-experimental study.

Chapter 3: Scales of Measurement

1. If participants name their county of residence, the resulting data are at what level?
   A. Ordinal.
   B. Interval.
   C. Ratio.
   D. Nominal.
2. If a teacher ranks students from low to high on their volleyball skills, he or she is measuring at what level?
   A. Ordinal.
   B. Interval.
   C. Ratio.
   D. Nominal.
3. Which two scales of measurement tell us by *how much* participants differ from each other?
   A. Ordinal and nominal.

   B. Interval and ordinal.

   C. Ratio and interval.

   D. Nominal and interval.

4. "The ordinal scale is a higher level of measurement than the interval scale." This statement is

   A. true.

   B. false.

5. "Measuring height using a tape measure is an example of the ratio scale of measurement." This statement is

   A. true.

   B. false.

Chapter 4: Descriptive, Correlational, and Inferential Statistics

1. Which of the following is used to summarize data?

   A. Inferential statistics.

   B. Descriptive statistics.

2. If there is a perfect correlation, what is the value of the correlation coefficient?

   A. 0.00.

   B. 1.00.

   C. Some other value.

3. "It is necessary to use inferential statistics when conducting a census." This statement is

   A. true.

   B. false.

4. "All populations are large." This statement is

   A. true.

   B. false.

5. Which type of statistics tells researchers how much confidence they can have when they generalize from samples to populations?

   A. Inferential.

   B. Descriptive.

Chapter 5: Introduction to Sampling

1. *Parameters* are based on a study of a

   A. sample.

   B. population.

2. "Using volunteers when sampling is presumed to create a bias." This statement is
   A. true.
   B. false.
3. What is the most important characteristic of a good sample?
   A. Being free from bias.
   B. Being large.
4. "Random sampling creates sampling errors." This statement is
   A. true.
   B. false.
5. Using random sampling identifies
   A. an accidental sample.
   B. a sample of convenience.
   C. an unbiased sample.

Chapter 6: Random Sampling

1. Putting the names of girls in one hat and those of boys in another hat and drawing 20% of the girls' names and 20% of the boys' names from each hat constitutes
   A. cluster sampling.
   B. stratified random sampling.
   C. simple random sampling.
2. Which of the following usually creates less sampling error?
   A. Simple random sampling.
   B. Stratified random sampling.
3. "Using stratified random sampling eliminates all sampling errors." This statement is
   A. true.
   B. false.
4. Suppose there are 500 people in a population and you want to draw a sample using a table of random numbers. Which of the following would be an appropriate number name for the first person to whom you assign a number before using the table?
   A. 00.
   B. 01.
   C. 05.
   D. 001.
5. Suppose a researcher randomly selected ten classrooms (as clusters). Each classroom had 20 students. The researcher should report the sample size as
   A. 10.

B. 20.

C. 200.

Chapter 7: Sample Size

1. Increasing sample size
   A. increases precision.
   B. decreases bias.
   C. increases precision *and* decreases bias.
2. Which of the following will produce a greater reduction in sampling errors?
   A. Increasing the size of a sample from 800 to 900 (an increase of 100).
   B. Increasing the size of a sample from 200 to 300 (an increase of 100).
3. "The smaller the anticipated difference between groups, the larger the sample size should be." This statement is
   A. true.
   B. false.
4. "For samples with very limited variability, even small samples can yield precise results." This statement is
   A. true.
   B. false.
5. "It is usually better to use a small, unbiased sample than a large, biased one." This statement is
   A. true.
   B. false.

Chapter 8: Standard Error of the Mean and Central Limit Theorem

1. "Using random sampling guarantees freedom from sampling errors." This statement is
   A. true.
   B. false.
2. "According to the central limit theorem, the sampling distribution of means is skewed." This statement is
   A. true.
   B. false.
3. The larger the variability in a population, the
   A. larger the standard error of the mean.
   B. smaller the standard error of the mean.

4. If $m = 40.00$ and $SE_M = 3.00$, what are the limits of the 68% confidence interval for the mean?
   A. 37.00 and 40.00.
   B. 40.00 and 43.00.
   C. 37.00 and 43.00.
   D. Some other values.
5. If you increase the sample size, what effect does this have on the size of the standard error of the mean?
   A. It increases it.
   B. It decreases it.

Chapter 9: Frequencies, Percentages, and Proportions

1. "In descriptive statistics, the lowercase letter f stands for *function*." This statement is
   A. true.
   B. false.
2. If there are 1,000 citizens in a town and 53% favor capital punishment, how many favor it?
   A. 53.
   B. 530.
   C. Some other number.
3. If 30 out of 100 parents favor school uniforms, what percentage favors them?
   A. 30%.
   B. 60%.
   C. 90%.
   D. Some other percentage.
4. "For a proportion of .22, the corresponding percentage is 2.2%." This statement is A   true. B   false.
   A. true.
   B. false.
5. "When reporting percentages, it is desirable to also report the underlying frequencies." This statement is
   A. true.
   B. false.

Chapter 10: Shapes of Distributions

1. "A frequency polygon is a drawing that shows how many participants have each score." This statement is

A. true.

B. false.

2. "A normal curve is also called a skewed curve." This statement is

A. true.

B. false.

3. When a curve has a tail to the left but no tail to the right, it is said to have a

A. positive skew.

B. negative skew.

4. "Income in large populations is usually skewed to the right." This statement is

A. true.

B. false.

5. "Another name for a skewed curve is 'bell-shaped curve.'" This statement is

A. true.

B. false.

Chapter 11: The Mean: An Average

1. "The uppercase letter X without a bar over it is a symbol for the mean." This statement is

A. true.

B. false.

2. "The mean is the most frequently used average." This statement is

A. true.

B. false.

3. "In a set of scores, the deviations from the mean have a sum of zero." This statement is

A. true.

B. false.

4. The mean is associated with which scales of measurement?

A. Ordinal and nominal.

B. Interval and ordinal.

C. Nominal and interval.

D. Ratio and interval.

5. "The mean is an especially good average for describing skewed distributions." This statement is

A. true.

B. false.

Chapter 12: Mean, Median, and Mode

1. Which average is defined as the *most frequently occurring score*?
   A. Mean.
   B. Median.
   C. Mode.

2. If the median for a set of scores equals 75, what percentage of the scores is below 75?
   A. 25%.
   B. 50%.
   C. 100%.
   D. Some other percentage.

3. "The mean is insensitive to extreme scores." This statement is
   A. true.
   B. false.

4. "In a distribution with a negative skew, the median has a higher value than the mean." This statement is
   A. true.
   B. false.

5. What is the mode of the following scores: 1, 2, 3, 6, 6, 6
   A. 3.
   B. 4.
   C. 6.
   D. Some other value.

Chapter 13: Range and Interquartile Range

1. "A synonym for the term *variability* is *dispersion*." This statement is
   A. true.
   B. false.

2. "The *range* is a statistic that describes central tendency." This statement is
   A. true.
   B. false.

3. Scores that lie far outside the range of the vast majority of scores are known as
   A. *IQR*s.
   B. outliers.
   C. median points.

4. "The interquartile range is defined as the range of the middle 50% of the participants." This statement is
   A. true.
   B. false.
5. "The interquartile range is unduly affected by outliers." This statement is
   A. true.
   B. false.

Chapter 14: Standard Deviation

1. "The standard deviation is a frequently used measure of variability." This statement is
   A. true.
   B. false.
2. Which group has a larger standard deviation?
   A. Scores for Group I: 0, 5, 10, 15, 20.
   B. Scores for Group II: 1, 2, 3, 4, 5.
3. "If all participants have the same score, the value of the standard deviation is 1.00." This statement is
   A. true.
   B. false.
4. In a normal distribution, what percentage of the cases lies between the mean and one standard deviation unit above the mean?
   A. 34%.
   B. 50%.
   C. 68%.
5. In a normal distribution with a mean of 50.00 and a standard deviation of 8.00, what percentage of the cases lies between scores of 42 and 58?
   A. 34%.
   B. 50%.
   C. 68%.

Chapter 15: z Score

1. When scores are standardized by units of standard deviation, these scores are called
   A. z scores.
   B. percentile scores.

2. Given the following *z* scores, which score shows one standard deviation below the mean?
   A. *z* score of 1.
   B. *z* score of –2.
   C. *z* score of –1.

3. Among the *z* scores given in the question above, which score would show the lowest percentile rank?

4. Given the normal distribution of the areas under the curve, what is the percentile rank of the following?
   A. *z* score of 2.
   B. *z* score of 3.

5. When working with a population mean and population standard deviation it is correct to subtract the population mean from the raw score, then divide it by the population standard deviation. This statement is
   A. true.
   B. false.

Chapter 16: Correlation

1. For the scores on Test X and Test Y shown below, there is
   A. a direct relationship.
   B. an inverse relationship.
   C. no relationship.

Student	Test X	Test Y
Janice	25	9
Brittany	30	7
Ramon	35	4
Wallace	40	1

2. For the scores on Test D and Test E shown below, there is
   A. a direct relationship.
   B. an inverse relationship.
   C. no relationship.

Student	Test D	Test E
Buddy	303	20
Turner	343	53
Kathy	479	70
Suzanne	599	88

3. "A *direct* relationship is sometimes called a *positive* relationship." This statement is
   A. true.
   B. false.
4. "In an inverse relationship, those who are high on one variable tend to be low on the other." This statement is
   A. true.
   B. false.
5. "*Correlation* is the best way to examine cause-and-effect." This statement is
   A. true.
   B. false.

Chapter 17: Pearson *r*

1. When there is a perfect, inverse relationship, what is the value of *r*?
   A. 1.00.
   B. 0.00.
   C. –1.00.
   D. Some other value.
2. "It is possible for a relationship to be both inverse and strong." This statement is
   A. true.
   B. false.
3. Which of the following values of *r* represents the strongest relationship?
   A. 0.64.
   B. –0.79.
   C. 0.00.
4. "An *r* of 0.60 is equivalent to 60%." This statement is
   A. true.
   B. false.
5. "An *r* of –0.95 represents a stronger relationship than an *r* of 0.88." This statement is
   A. true.
   B. false.

Chapter 18: Scattergram

1. Scattergrams are presented in research reports
   A. frequently.
   B. infrequently.

2. The more scatter in a scattergram, the
   A. stronger the relationship.
   B. weaker the relationship.
3. "In the social and behavioral sciences, it is common to find scattergrams in which all the dots are on a single straight line." This statement is
   A. true.
   B. false.
4. "When the dots in a scattergram form a pattern from the lower left to the upper right, the relationship is inverse." This statement is
   A. true.
   B. false.
5. "Each dot in a scattergram stands for the two scores of one participant." This statement is
   A. true.
   B. false.

Chapter 19: Coefficient of Determination

1. "The symbol for the coefficient of determination is r^2." This statement is
   A. true.
   B. false.
2. If the Pearson r equals 0.30, the coefficient of determination is calculated by
   A. taking the square root of 0.30.
   B. multiplying 0.30 by 0.30.
3. "For an r of 0.80, the ability to predict is 64% better than zero." This statement is
   A. true.
   B. false.
4. When a coefficient of determination equals 0.20, what percentage of the variance on one variable is *not* predicted by the other variable?
   A. 4%.
   B. 20%.
   C. 96%.
   D. Some other percentage.
5. "When $r = 0.40$, the percentage of variance accounted for is 16%." This statement is
   A. true.
   B. false.

Chapter 20: Multiple Correlation

1. When low scores on one variable are associated with low scores on the other variable, this suggests that the relationship is
   A. direct.
   B. inverse.
2. Which of the following values of R represents the weakest relationship?
   A. $R^2 = 1.00$.
   B. $R^2 = -0.96$.
   C. $R^2 = 0.22$.
3. For determining the correlation between one variable as a predictor of a second variable, which of the following should be computed?
   A. R.
   B. r.
4. If $R^2 = 0.75$, what percentage of the variance is accounted for?
   A. 25%.
   B. 75%.
5. Suppose a researcher is examining the validity of a combination of scores on a spatial relations test and previous math grades as a predictor of geometry grades. Which correlational statistic should the researcher compute for this research problem?
   A. R.
   B. r.

Chapter 21: Introduction to Hypothesis Testing

1. Which of the following is a correct statement of the null hypothesis?
   A. There is a true difference between the means.
   B. There is no true difference between the means.
2. Which of the following is a symbol for the null hypothesis?
   A. H_0.
   B. H_1.
3. "For a given study, the research hypothesis and the null hypothesis usually say the same thing." This statement is
   A. true.
   B. false.
4. Which type of hypothesis predicts that one particular group's mean will be higher than another group's mean?
   A. Directional hypothesis.
   B. Non-directional hypothesis.
   C. Null hypothesis.

5. The null hypothesis states that the true difference between the means
   A. equals zero.
   B. is greater than zero.
   C. is less than zero.

Chapter 22: Decisions about the Null Hypothesis

1. Which of the following yields a probability?
   A. A descriptive statistic.
   B. A significance test.
2. At what point is it conventional to reject the null hypothesis?
   A. When the probability is less than 0.05.
   B. When the probability is greater than 0.05.
3. The null hypothesis can be rejected with the greatest confidence when which one of the following is true?
   A. $p < 0.05$.
   B. $p < 0.01$.
   C. $p < 0.001$.
4. Rejecting the null hypothesis when in reality it is true is known as a
   A. type I error.
   B. type II error.
5. By conventional standards, if $p < 0.01$, researchers declare the difference statistically
   A. insignificant.
   B. significant.

Chapter 23: Limitations and Implications of Significance Testing

1. "For the difference between two means, the null hypothesis says that the difference is large." This statement is
   A. true.
   B. false.
2. "Equating a significant difference with a large difference is a mistake." This statement is A   true. B   false.
   A. true.
   B. false.

3. Is the size of the sample one of the factors that contributes to determining statistical significance?
   A. Yes.
   B. No.
4. Can a very small difference between two means be statistically significant?
   A. Yes.
   B. No.
5. Evaluating a difference is a three-step process. What is the first step?
   A. Determining whether it is statistically significant.
   B. Considering the practical implications of an insignificant difference.
   C. Considering the practical implications of a statistically significant difference.

Chapter 24: Introduction to the *t* Test

1. "A *t* test yields a probability." This statement is
   A. true.
   B. false.
2. "The smaller the sample, the more likely that the null hypothesis will be rejected." This statement is
   A. true.
   B. false.
3. Under which of the following circumstances is the null hypothesis more likely to be rejected?
   A. When there is a small observed difference between means.
   B. When there is a large observed difference between means.
4. "Dependent data may have less sampling error than independent data." This statement is
   A. true.
   B. false.
5. If participants are matched (i.e., paired) across experimental and control groups, the resulting data are
   A. independent.
   B. dependent.

Chapter 25: Independent Samples *t* Test

1. Groups like experimental and control groups, athletes and non-athletes, or alcoholics and non-alcoholics are examples of a categorical variable that is used in an Independent Samples *t* Test as the
   A. independent variable.
   B. dependent variable.

2. When a null hypothesis is rejected, we can conclude that
   A. the mean difference was statistically not significant.
   B. the mean difference was statistically significant.
3. If the first sample of an n of 50 and the second sample of an n of 45 are being compared in an Independent Samples t Test, the degrees of freedom (df) would be?
   A. 95.
   B. 93.
4. Which of the following states a directional hypothesis?
   A. $\mu_1 - \mu_2 = 0$.
   B. $\mu_1 \neq \mu_2$.
   C. $\mu_1 > \mu_2$.
   D. $\mu_1 = \mu_2$.
5. The dependent variable is a variable that depends on or is the outcom.e of the independent variable. This statement is
   A. true.
   B. false.

Chapter 26: Dependent Samples *t* Test

1. The purpose of a Dependent Samples t Test is to compare the difference between two
   A. related or dependent samples.
   B. independent samples.
2. The Dependent Samples t Test uses one categorical data and one continuous data, and the categorical data is the
   A. independent variable.
   B. dependent variable.
3. The Dependent Samples t Test assumes what about the observations?
   A. They are dependent.
   B. They are independent.
4. The degrees of freedom for the Dependent Samples t Test is
   A. $df = N - 1$.
   B. $df = n_1 + n_2$.
5. When the results indicate that paired samples analysis produced a significant t value [$t\,(99) = 15.14$, $p < 0.001$], this means that the means are
   A. significantly different.
   B. not different.

Chapter 27: One-Sample *t* Test

1. The purpose of a One-Sample *t* Test is to compare the difference between two
   A. related or dependent samples.
   B. independent means.
2. The One-Sample *t* Test uses continuous data that is normally distributed. This variable is the
   A. independent variable.
   B. dependent variable.
3. The degrees of freedom for the One-Sample *t* Test is
   A. $df = N - 1$.
   B. $df = n_1 + n_2$.
4. When the results indicate that a One-Sample *t* Test produced a significant *t* value [$t(42) = 5.59, p < 0.01$], this means that the means being compared are
   A. significantly different.
   B. not different.
5. Calculating the *t* value of a One-Sample *t* Test involves taking the difference between the sample mean and the population mean relative to the standard error of mean. This statement is
   A. true.
   B. false.

Chapter 28: Reporting the Results of *t* Tests: Display of Outcomes

1. "Reporting a *t* test makes it unnecessary to report the values of the means and standard deviations." This statement is
   A. true.
   B. false.
2. If you read that $t = 0.452, df = 100, p > 0.05$, what should you conclude?
   A. the difference is statistically significant.
   B. the difference is not statistically significant.
3. "If a *t* test yields $p < 0.05$, the null hypothesis normally would be rejected." This statement is
   A. true.
   B. false.
4. If a researcher rejects the null hypothesis, what else is true?
   A. the difference is statistically significant.
   B. the difference is not statistically significant.

5. "It is safe to assume that if a difference is statistically significant, it is of practical significance." This statement is
   A. true.
   B. false.

Chapter 29: One-Way ANOVA

1. "ANOVA can be used to test for the difference(s) between only two means." This statement is
   A. true.
   B. false.
2. "The acronym ANOVA stands for *analysis of variance*." This statement is
   A. true.
   B. false.
3. For the typical consumer of research, which one of the following values in an ANOVA table is of greatest interest?
   A. mean squares.
   B. the value of p.
   C. the value of F.
4. Suppose you read the following: $F = 0.641$, $df = 3, 29$, $p > 0.05$. What conclusion would you normally draw about the null hypothesis?
   A. reject it.
   B. do not reject it.
5. Suppose you read the following: $F = 3.50$, $df = 2, 20$, $p < 0.05$. What conclusion would you normally draw about statistical significance?
   A. it is statistically significant.
   B. it is not statistically significant.

Chapter 30: Two-Way ANOVA

1. Suppose participants were classified according to their religion and their country of origin in order that means for both religious groups and national origin groups be compared. This would call for a
   A. one-way ANOVA.
   B. Two-Way ANOVA.
2. In order to examine an *interaction*, you
   A. temporarily ignore one way that the participants were classified while examining the results of the other way they were classified.

    B. look at both ways participants were classified at the same time in order to see how the two classification variables affect each other.

3. "In the table below, there appears to be an interaction." This statement is
    A. true.
    B. false.

	X	Y
A	$M = 40.00$	$M = 30.00$
B	$M = 30.00$	$M = 40.00$

4. "In the table below, there appears to be an interaction." This statement is
    A. true.
    B. false.

	X	Y
A	$M = 300.00$	$M = 200.00$
B	$M = 350.00$	$M = 250.00$

5. "In the table below, there appear to be two main effects." This statement is
    A. true.
    B. false.

	X	Y
A	$M = 40.00$	$M = 50.00$
B	$M = 30.00$	$M = 20.00$

Chapter 31: Chi-Square Test

1. "For nominal data, researchers normally report frequencies and percentages instead of means and standard deviations." This statement is
    A. true.
    B. false.

2. The symbol for Chi-Square is
    A. p.
    B. r^2.
    C. χ^2.
    D. p^2.

3. "For the data in the following table, a Two-Way Chi-Square would be an appropriate test of significance." This statement is
    A. true.
    B. false.

	Happy	Unhappy
Boys	$n = 30$	$n = 40$
Girls	$n = 40$	$n = 30$

4. Suppose you read that as the result of a Chi-Square Test, $p < 0.001$. By conventional standards, what decision should be made about the null hypothesis?
   A. Reject it.
   B. Do not reject it.

5. Suppose you read that as the result of a Chi-Square Test, $p < 0.05$. By conventional standards, what decision should be made about statistical significance?
   A. It is significant.
   B. It is not significant.

Chapter 32: Effect Size

1. "The purpose of *effect size* is to standardize the difference between two means." This statement is
   A. true.
   B. false.

2. To compute Cohen's *d*, first subtract the control group's mean from the experimental group's mean, and then do what?
   A. multiply the difference between the means by the control group's standard deviation.
   B. divide the mean difference by the square root of the pooled sample variance.
   C. determine the significance of the difference between the means at $p < 0.05$.

3. What does it mean when $d = 1.00$?
   A. the average participant in the experimental group is one standard deviation higher than the average participant in the control group.
   B. the average participant in the control group is one standard deviation higher than the average participant in the experimental group.

4. "For most practical purposes, values of *d* rarely exceed 3.0." This statement is
   A. true.
   B. false.

5. "A large effect size might represent an unimportant result." This statement is
   A. true.
   B. false.

Chapter 33: Simple and Multiple Linear Regression

1. If a research has one categorical variable predicting one continuous variable, this is an example of
   A. simple linear random.
   B. multiple regression.

2. If a researcher has one continuous variable predicting another continuous variable this is an example of
   A. simple linear random.
   B. multiple regression.

3. A teacher found that the phonics method of teaching reading resulted in an increase in students' reading scores. The phonics method in this study is the
   A. independent variable.
   B. dependent variable.

4. Research investigating the effect of parental involvement on the academic achievement of their children found that reading with children at home predicted high scores in the children's language arts test scores. Reading with children in this study is the
   A. independent variable.
   B. dependent variable.

5. Between the X and Y symbols, X represents the _____ and Y represents the _____.
   A. independent variable, dependent variable.
   B. dependent variable, independent variable.

Glossary

Introduction

Quantitative research: uses numerical data to determine a cause-effect relationship (Intro)

Qualitative research: uses verbal narrative data to describe how things are and what they mean (Introduction)

Research: coming to know something using a scientific method, which is a systematic way of acquiring new knowledge (Introduction)

Scientific method: a method of gaining knowledge through: 1. Identifying a problem. 2. Formulating hypotheses. 3. Determining what information needs to be collected. 4. Organizing the gathered information. 5. Interpreting the results (Introduction)

Part A

Chapter 1

Empiricism: uses direct observation to obtain knowledge (1)

Empirical approach: acquiring knowledge based on making observations of individuals or objects of interest (1)

Empirical research: use of the empirical approach to acquire knowledge (1)

Everyday observations: an application of the empirical approach (1)

Examinees: individuals who take an examination such as an achievement test (1)

Hypothesis: a prediction of the outcome of research (1)

Observations: collecting information (1)

Participants: individuals being studied who have freely consented to participate in the research (1)

Population: population consists of all members of the groups of interest to one's research study (1, 5, 6)

Respondents: individuals who respond to a survey such as a political poll (1)

Sample: a subset of a population (1)

Subjects: those being studied have not consented to participating such as animals or individuals observed unobtrusively without their consent (1)

Chapter 2

Cause-and-effect relationships: the independent variable is the possible cause and the dependent variable demonstrates the possible effect (2)

Control group: the group receives no control condition (2)

Dependent variable: the responses to the treatment in an experiment, the predicted or outcome, represented by the symbol, Y (2, 22, 33)

Experimental group: the group receives a control condition (2)

Experimental study: a design in which treatments are given to see how the participants respond to them (2)

Independent variable: treatment in an experiment, a predictor(s) represented by the symbol, X, to predict the dependent variable, Y (2, 22, 33)

Non-experimental study/design: (also called a descriptive study): is a study in which observations are made to determine the status of what exists at a given point in time without the administering of treatments (2)

Program evaluation: conducted to ascertain if a program in place is effective for a specific group. The goal of a program evaluation is to measure the quality, merit, and worth of the program's goals (12)

Chapter 3

Categorical data: nominal and ordinal levels of measurements are known as categorical data (3)

Dichotomous variable: a variable that has two categories that are mutually exclusive between the two categories, but inclusive of all participants in the two categories is known as a dichotomous variable

Interval scales: measures how much participants differ from each other. The scale does not have an absolute zero point on its scale (3)

Nominal: the lowest level of measurement. It is the naming level because names (i.e., words) are used instead of numbers (3)

Numerical (or continuous) data: interval and ratio levels of measurements. These scales are expressed in numbers unlike categorical variables (3)

Ordinal: participants are ranked in order from low to high, but it does not indicate how much lower or higher one participant is in relation to another (3)

Ratio scales: measure how much participants differ from each other. The scale has an absolute zero point on its scale (3)

Scales of measurement: types of variables: nominal, ordinal, interval, and ratio (3)

Variables: any measurable factor that has an effect on a phenomenon or phenomena (3)

Chapter 4

Average: is a descriptive statistic (4)

Census: a study in which all members of a population are included (4)

Correlational statistics: are a special subgroup of descriptive statistics. The purpose of correlational statistics is to describe the relationship between two or more variables for one group of participants (4)

Correlation coefficient: range in value from 0.00 (no correlation between variables) to 1.00 (a perfect correlation)

Descriptive statistics: summarize data (4)

Inferential statistics: tell us how much confidence we can have when generalizing from a sample to a population (4)

Margin of error: an inferential statistic that measures to what degree the result from the sample is inaccurate as caused by the sample representing the population It is reported as a warning to readers of research that random sampling may have produced errors, which should be considered when interpreting results (4, 8)

Percentage: the symbol is %, indicates the number per 100 who have a certain characteristic (4, 9)

Range: scores from the highest to the lowest, which would indicate how much the scores vary (4)

Part B

Chapter 5

Biased sample: when some members of a population have a greater chance of being selected than others (5)

Parameters: summarized results (such as averages) of the population (5)

Population: see Chapters 1, 5, and 6

Random sampling: eliminates bias in the selection of individuals for a study (5, 6)

Sample: a subset of a population (1, 5)

Sample of convenience (accidental samples): increases the odds of some members of a population being selected, creating bias (5)

Sampling errors (random errors or random chance errors): sampling errors measure the degree of misrepresentation of the sample from the population. Both biased and unbiased random samples still contain random sampling errors (5, 6, 7, 8, 19)

Simple random sampling: a researcher can put names on slips of paper and draw the number needed for the sample. This method is efficient for drawing samples from small populations. Each member of a population is given an equal chance of being selected (5, 6)

Statistics: summarized results of a sample (5)

Unbiased sample: individuals have an equal chance of being selected (5)

Volunteerism (self-selection bias): comprise the sample with those who self-select excluding those who decide not to participate, thereby creating sampling bias (5)

Chapter 6

Multistage random sampling: used for large-scale studies using several stages to draw the sample (6)

Population: see Chapters 1, 5, and 6

Proportionate stratified sampling: the same proportions of subgroups in the population is retained in the sample (6)

Random cluster sampling: existing groups of participants are drawn (6)

Simple random sampling: researcher puts names on slips of paper and draws the number needed for the sample (6)

Stratified random sampling: the population is first divided into strata that are believed to be relevant to the variable(s) being studied, then simple random sampling is used to select the participants (6)

Table of random numbers: give each person in the population a *number name*, used to draw a random sample (6)

Chapter 7

Diminishing returns: an increase in sample size to an already larger sample size would give the researcher **diminishing returns** (7)

Effect size (d) = the estimated difference between two means divided by pooled estimated standard deviation (7, 32)

Power analysis helps determine what sample size we need to correctly conclude the significant difference (7)

Precision: the extent to which the same results would be obtained with another random sample from the same population (7)

Chapter 8

Central limit theorem: states that the *sampling distribution of means* is normal in shape. The central limit theorem makes it possible to estimate the standard error of the mean given only one mean and the associated standard deviation (8)

Standard error of the mean (SE_M): the standard deviation of the sampling distribution (8, 24)

The 68% confidence interval for a mean: about 68% of all sample means lie within one standard deviation unit of the mean (8)

Part C

Chapter 9

Frequency: is the number of participants or cases; its symbol is f. The symbol for the number of participants is *N* (9)

Proportion: is part of one (1) and usually a decimal number or fraction (9)

Chapter 10

Bimodal distributions: have two high points (10)

Curve: when there are many participants, thus more numbers, the shape of a polygon becomes smoother and is referred to as a curve (10)

Frequency distribution: a table that shows how many participants have each score (10)

Frequency polygon: a drawing that shows how many participants have each score (10)

Negative skew: a distribution that has a long tail is pointing to the left (i.e., skewed to the left or the negative side) (10)

Normal curve (bell-shaped curve): curve is low on the left and right sides and high in the middle. It is used as the basis for a number of inferential statistics (10)

Positive skew: a distribution that has a long tail pointing to the right (i.e., skewed to the right or the positive side) (10)

Chapter 11

Deviations: calculate how far each score is away (or deviates) from the mean (11)

Mean: the most frequently used average. M and m are the most commonly used symbols for the *mean* in academic journals. It is defined as the balance point in a distribution of score or the balance point in a distribution of scores (11, 12)

Chapter 12

Median: the value in a distribution that has 50% of the cases above it and 50% of the cases below it. It is defined as the middle point in a distribution (12)

Mode: the most frequently occurring score (12)

Mean: see Chapter 11 (12)

Chapter 13

Variability: differences among the scores of participants—how much scores vary from each other (13)

Measures of variability: concisely describe the amount of variability in a set of scores (13)

Range: the difference between the highest score and the lowest score (13)

Outliers: score that lies far outside the range of the vast majority of other scores and increases the size of the range (13)

Interquartile range (IQR): the range of the middle 50% of the participants (13)

Chapter 14

Standard deviation: the most frequently used *measure of variability*. *S* is the symbol for the *standard deviation* (14)

Variability (*spread or dispersion*): refers to the differences among the scores of participants (14)

The 68% range: (sometimes called *the two-thirds rule of thumb*) About two-thirds of the cases lie within *one standard deviation unit* of the *mean* in a normal distribution (14)

Chapter 15

z scores: standardizes the raw scores in units of standard deviation. Z scores help us understand where the scores fall in relation to the other scores in the data (15)

Percentile scores: refers to the percentage of the area that falls below a given score (15)

Part D

Chapter 16

Correlation: refers to the extent to which two variables are related across a group of participants (16, 17, 18)

Direct (**or positive relationship**): those who score high on one variable tend to score high on the other, *and* those who score low on one variable tend to score low on the other (16)

Inverse (**or negative relationship**): those who score high on one variable tend to score low on the other (16)

Chapter 17

Pearson r: (symbol *r*, also known as *Pearson product-moment correlation coefficient*, *Pearson correlation coefficient*, or the *product-moment correlation coefficient*) statistic for describing the relationship between two variables, ranges from −1.00 to 1.00 (17)

Test reliability: is to administer the same test twice to a group of participants without trying to change the participants between administrations of the test. If *r* is high, the test results are consistent and the test is said to be reliable (17)

Chapter 18

Correlation: see Chapter 16

Scattergram: (also known as a *scatter diagram*) illustrates correlation between two variables (18)

Coefficient of determination: r^2 (symbol,) when converted to a percentage, indicates how much variance in one variable is accounted for by the variance in the other (19, 20)

Chapter 19

Coefficient of determination: (symbol, r^2) when converted to a percentage, indicates how much variance on one variable is accounted for by a combination of predictors (18, 20)

Chapter 20

Multiple correlation coefficient: (symbol, R) used to determine the degree of relationship between a combination of the two predictors (32)

Part E

Chapter 21

Alternative hypothesis (H_1): an alternative to the null hypothesis (21)
Directional research hypothesis (also called a *one-tailed hypothesis*): states that one particular group's average is higher than the other group's (12)
Non-directional research hypothesis: (also called a *one-tailed hypothesis*): a hypothesis that there is a difference (that the populations of the two groups are not equal) but he or she is not willing to speculate on the direction of the difference (21)
Null hypothesis (H_0): says that the true difference between the means is zero. The null hypothesis is a possible explanation for an observed difference when there are sampling errors (21)
Research hypothesis: a researcher's expectation (21)
Sampling errors: see Chapter 5 (21)

Chapter 22

P value (p): the probability that the null hypothesis is true (22)
Type I error: the error of rejecting the null hypothesis when it is correct (22)
Type II error: the error of failing to reject the null hypothesis when it is false (22)

Chapter 23

Practical implications: *insignificant* differences can have practical implications and significant differences may not have practical implications (29)
Statistically significant: (reliable) when a null hypothesis has been rejected, a researcher declares the difference in what's being tested (29)

Part F

Chapter 24

Degrees of freedom: count the number of independent pieces of information used for the result (N) minus the number of groups (24)

The t test: tests the difference between two sample means to determine statistical significance (24)

Independent Samples t Test: compares two means of *independent data* (sometimes called *uncorrelated data*) (24, 25)

Dependent or paired-Samples t test: for *dependent data* (sometimes called *correlated data*) (24, 26)

One-Sample t Test: compares a sample mean to a population mean or a historical mean (24, 27)

Chapter 25

Dependent variable: a variable that depends on or is the outcome of the independent variable or the predictor (25)

Independent Samples t Test: compares the difference between two independent sample means (also see Chapter 24) (25)

Independent variable: the variable that may affect the dependent or the outcome variable (25)

Standard error of the difference between means: measures the average amount of error from the two sample means using the standard error of each sample (25)

Chapter 26

Dependent Samples t Test: comparison of two means from one sample with two repeated measures or related samples (also see Chapters 24 and 25) (26)

Within-subject design (repeated-measures design): One sample with repeated measures for a Dependent Samples *t* Test (24, 26)

Chapter 27

One-Sample t Test: see Chapter 24

Standard error of mean: see Chapter 8

Chapter 28

Statistically significant: synonymous with *rejecting the null hypothesis*, the phrase used to compare means in inferential statistics (25)

Chapter 29

Analysis of variance: ANOVA: used to test the difference(s) among two or more means (29)

ANOVA table: sometimes used to report the results (29)

Effect size: measures the percentage of the unknown in one variable that is explained by another variable (29, 32)

Eta squared: (η^2) measures how much of the variable is explained by group differences (29)

One-Way ANOVA (also known as a single-factor ANOVA): each participant is classified in one way with multiple categories to compare (29)

Post-hoc tests (after-the-fact tests): determine which specific difference(s) is (are) significant. Types of post-hoc tests are Tukey's HSD test and Scheffé's test among others (29)

Tukey's HSD test and Scheffé's test: types of *post-hoc tests* (after-the-fact tests) to compare group means

Two-Way ANOVA (also known as a two-factor ANOVA): in which each participant is classified in two ways, each with multiple categories to compare. It examines two *main effects* and one *interaction* (29, 30)

Chapter 30

Main effect: the result of comparing one of the ways in which the participants were classified while temporarily ignoring the other way in which they were classified (30)

Interaction: when the effect of one grouping variable on the dependent variable is affected by the other grouping variable (30)

Part G

Chapter 31

Chi-Square Test of Independence (χ^2): used when the data are from two categorical or nominal variables (31)

Non-parametric statistics: statistical tests with the violation of the assumption of normality (31)

One-Way Chi-Square: (also known as a *goodness of fit Chi-Square Test*) participants are classified in only one way (31)

Parametric tests: statistical tests with the assumption that the underlying distributions are normal (31)

Two-Way Chi-Square Test: the participant is classified according to two ways (31)

Chapter 32

Cohen's d: used to measure effect size. Calculation: subtract the control group mean from the experimental group mean and divide by square root of pooled sample variance (30)

Effect size: standardizes the size of the difference between two means in the control group's standard deviation units (32)

Chapter 33

Dependent variable: see Chapters 2 and 25 (33)

Independent variable: see Chapters 2 and 25 (33)

Multiple regression: used to measure the predictability of more than one independent variable (categorical or continuous) and one dependent variable (continuous) (33)

Simple linear regression: a common tool in the social sciences to measure predictability using one independent variable (continuous or categorical) and one dependent variable (continuous) (33)

Index

Page numbers in *italics* mark figures, **bold** indicate tables, and "n"/"nn" refer to notes.

absolute zero 17, **18**
accidental samples 28–29
alpha level 146n2
alternative hypothesis (H_A) 135–36, 138–40, 164; symbolic statement 165, 171, 176
ANOVA, analysis of variance 186; and effect size 212, **212**, 213; and the null hypothesis 187; one-way or single factor 185, 188, **189**; and probability 143; tables 187, 201n3; two-way or two-factor 189, **189**, 193–94, 198–99, *see also F* tests
average scores 21
averages 69, 71–72, *77*, *see also* mean; median; mode

B coefficient 222n3
ß coefficient 222n3
bell-shaped curve 64–65, *76*, *see also* curves; normal curve
bias 28, 33, 41, 44–45, 53–54
biased samples 28–30, 45
bimodal distributions *67*

categorical data 16, **18**
causal relationships 101
cause-and-effect relationships 10, 13n1, 101
census 23
central limit theorem 50
cheerfulness/depression examples 5, 64, **100**, 101, 116, 117, *118*
Chi-Square Tests of Independence (χ^2) 206–7; one way or goodness of fit 207; results reporting 209; two way 207–8, 209n3
coefficient of determination (r^2) 111–12, *116*, 120n2, 121–22, **124**, 127; and effect size 212, **212**, 213, *see also* Pearson *r*

Cohen's *d*, and effect size 212–14, 228–29
College Entrance Test Scores example 106–7
conclusions, methods of coming to xi–xii
confidence intervals for the mean 52, **53**, 54
continuous data 176
control groups 11; standardizing 212
correlation 99, 111, 218
correlation coefficients 22, 24n1, *see also* multiple correlation coefficient *(R)*
correlation ratio 232
correlational statistics 22, 97
correlational studies 101, 102
curves *76*, *77*; and polygons 63–64, *see also* bell-shaped curve; normal curve

data analysis xii–xiii, 5
degrees of freedom *(df)* 159, 171; and Dependent Samples 171; and Independent Samples *t* tests 166; One Sample *t* test 176; and *t* test reports 180–82
dependent data 157–58
dependent *(or paired)* samples *t* tests 157–58, **159**, 167, 169–70, **171**, 228; and degrees of freedom 171
dependent variables 10, 164, 222n1; and linear regression analysis 218
descriptive statistics 21–22, 25, 57, 97, 152n3; averages 21–22; correlational statistics 22; range of scores 21–22
descriptive studies 9–10
deviations 70, **70**, 72n3, *see also* standard deviations
dichotomous variables 18
difference evaluation 148–49, 172
diminishing returns 42–43

direct (*or* positive) relationships 100–101, 112, *113*, *114*, 119, 130
directional research hypothesis 138, 176
dispersion 83n1

effect size 45, 187, 188, 211–13, **212**, 214–16; Cohen's *d* 45, 212, 213, **214**; formula 227–29; labeling **215**; principles of 215; and standard deviations 45; symbols for 213
empirical approach (*or* scientific method) 3–4
empirical research 4–5
empiricism 3
eta squared (η^2) 188
examinees 6
exceptions 102, 114–15
experimental groups 11
experimental studies 10
experiments 101; informal 10–12

F tests 185, 187, 191n2, 206, 210n2; vs. *t* tests 185–87, *see also* ANOVA, analysis of variance
frequencies (*f*) 59, **60**, 61, 61n1, 206
frequency distributions **18**, 63, **64**
frequency polygons 63, *64*, 65
frequency tables 91

generalizations 3–4
generalizing 33
Gosset, William 156

historical mean 175
hypothesis 3–4, *see also* alternative hypothesis (H_A); null hypothesis (H_0)
hypothesis testing 136

independent samples *t* tests 156–58, **159**, 161n5, 163–64; and degrees of freedom (*df*) 165, 166; formula 227; reporting 166
independent variables 10, 164; and linear regression analysis 218–19
inferential statistics 22–23, 24n2, 133, 139, 203, 205; and normal distributions 205–6; probability levels/statistical significance 143; and sampling errors 41, *see also* sampling errors
inferential tests 139, 141
interaction effects 195–98
interquartile range (IQR) *81*, 82, 83n3
interval estimates 53
interval scales 17, **18**, 19n1, 71, 75
inverse (*or* negative) relationships 101, *118*, 119
IQ tests example 208–9

Kruskal–Wallis *H* Test 168n1

main effects 194–95, 197–99
Mann-Whitney *U* Test 168n1

margin of error 22, 52, *see also* standard error of the mean (SE_M)
math scores example 112, *113*, 113, *115–16*, 117, 127–29
mathematical deduction 7n1
mean **18**, 50, **53**, 69–70, 73, 75, **76**, 78n3; column means 194, 198, 201n1; computation of 69–70; and deviations 70, **70**, 72n2, 86; population mean 50, 52–53; row means 194–95, 201n1; sample mean 176–77; and standard deviation 45, 52, 55n4, 87–89; symbols 70, 72n1; and *t* test reports 179, *see also* averages
mean and standard deviation table 179, **180–81**
means comparison 153, *see also t* tests
measurement scales, nominal 75–76
measures of central tendency 72, **76**, 78n1
measures of variability 80, **82**
median **18**, 73–76, *76*, 78n2, 82, 83n3, *see also* averages
medical research, and samples 30
mode **18**, 74–76, *76*, *see also* averages
multiple correlation coefficient (*R*) 127–30, *131*, *see also* correlation coefficients
multiple linear regression 217–18
multistage random sampling *36*

N or n symbol 61n2
naming level 16
negative relationship. *see* inverse relationships
negative skew distributions 66, 66, 76, *77*
NOIR 17
nominal data 75, 206–7
nominal measurements 16, **18**
non-experimental designs 9–10
non-experimental studies 13n1
non-parametric statistics 203, 205, 206
nondirectional research hypothesis 138, 176
normal curve *65*, 93–94; and sampling distribution of means 50, *51*; and standard deviations *87*, 88
normal distributions *65*, *76*, 205, 210n2, 216n3; and inferential statistics 205–6
null hypothesis (H_0) 135, 138–40, 147, 152n3, 156, 157, 175–76; and *p* values 142; rejecting 142–43, 147, 156–57, 160n1, 166–67, 180, 182, 183n2, 187, 199, 209; symbolic statement 137, 164, 170, 176; testing 198, *see also* significance tests
numerical (or continuous) data: interval scales 16, **18**, 19n1, 71, 75; ratio scales 16, 17, **18**, 19n2, 71

observations xi; direct 3, 5; everyday 3–4; indirect 5
observed differences 141, 146n1
one-sample *t* tests 157–59, **159**, 175, 228

one-way ANOVA 185, 188, **189**
one-way chi-square tests 207
opinion polls 43
ordinal measurements 16, **18**
outliers 81

p values 141–43, 148, 166, 182
paired (*or* dependent) samples *t* tests 157–58, **159**, 168, 169–70, **171**, 228; and degrees of freedom 171
parameters 28
parametric tests 176, 205
participants 6
Pearson product moment correlation coefficient 105
Pearson *r* **18**, 105–6, *106*, 107, 114–16, 125n1, 128, 231–32; formula 226–27; and variance 122, **124**, *see also* coefficient of determination (*r²*)
percentages 22, 59–60, **61**, 122, 206
percentile scores 93–94
perfect relationships 113, 119
phi **18**
point estimates 53
population mean 50, 138–39, 175
population parameters 28
populations 23, 27, 33, 61n2, 160n1, 176; and sampling 34–35, 44, 51, 157
positive relationships. *see* direct relationships
positive skew distributions 65, *66*, *77*
post hoc tests 188
power analysis 45
practical implications 150
precision 41–44, 51
probability (*p*) 142–43, 145–46, 171, 220
program evaluations 11–13
proportionate stratified sampling 35
proportions 60

qualitative research xii–xiii
quantitative research xii–xiii, 9, 21
questions, and research 4

random cluster sampling 36, *37*
random names exercise 122
random sampling 22, 23, 29–30, **37**, 49, 50, 137, 157, 206; cluster 36, *37*; multistage *36*; simple 34; stratified 34, 35, 47nn1–2; table of random numbers **34**, 39nn2–3, *see also* sample size; sampling; simple random sampling
range of scores 21–22
ranges 80–81, 83n2, *see also* interquartile range (IQR)
ratio scales 16, 17, **18**, 19n2, 71
regression analysis: and effect size **212**; and independent variables 218–20; multiple

linear 217–18; and the Pearson *r* 231–32; simple linear 217–19
rejecting the null hypothesis 142–44, 147, 156–57, 160n1, 166–67, 180–82, 184n2, 187, 199, 209
reliability 148–49
repeated-measures design 170
research studies xiii
research xi, 4–5; flawed 5
research hypothesis 137; directional 138; nondirectional 138
research methodologies, and effect size 215
research planning 4–5
respondents 6
responses. *see* dependent variables

sample means 176–77
sample sizes 30, 41–45, 51, 53
samples 27; biased 28–30; samples of convenience/accidental samples 28, *see also* random sampling
samples of convenience 28
sampling 4, 23, 25, 27–29, *see also* random sampling
sampling distribution of means 50, *51*
sampling errors 28–30, 35, 41, 49–50, 54, 137, 139, 147, 156–57, 206, *see also* inferential statistics
SAT-V, verbal portion of the *SAT* 99–100, 104n1
scales, standardizing 91–92
scales of measurement 15–17, **18**, 71, *see also* variable types
scatter diagrams 111
scattergrams 111, *113*, *115–16*, *118*, 119, 120
Scholastic Aptitude Test (SAT) 99
scientific method xii, 3
self-concept scores 100–101
self-selection bias 28
semi-interquartile range 81
sensory system xi
significance tests 141, 149–50
simple linear regression 217–19
simple random sampling 34, *see also* random sampling
single factor ANOVA. *see* one-way ANOVA
the 68% range 52–53, 87–88
skewed distributions 65, *66*, 66, 71, 75–76
Spearman *r* **18**
spread 83n1
standard deviations 18, 52, 85–87; calculating 88, 224–26; and effect size 45; formula 223–24; and the mean 45, 52, 55n4, 87–89; and the normal curve *87*, 88; symbols 86, 90n2; and z-scores 93–94, *see also* deviations; variability
standard error of the difference between means 165–66, 171

standard error of the mean (SE_M) 51–52, 177; formula 223, *see also* confidence intervals for the mean; margin of error
standardizing effect sizes 212, 213
statistical differences 165, 176
statistical significance 45, 136, 138, 142–43, 148–51, 152n2, 160n1, 180–81, 209, 211, 215, *see also t* tests
statistical software programs 166, 177
statistics 28
stratified random sampling 34–35, 47nn1–2
subjects 6
surveys 9–10

t tests 148, 155, 157, **159**, 160n3, 161n5, 191nn1–2, 206, 210n2; and ANOVA 186; and Cohen's *d* 213; dependent samples 228; and effect size **212**; and *F* tests 185–87; Independent Samples 156–57, 163–66, 227; one-sample 157–59, 175, 228; paired or dependent samples 157–58, 167, 169–70, **171**; reports 179–82, *see also* statistical significance
t values 171, 176, 177, 180, 220
table of random numbers **34**, 39nn2–3, 233–34
test reliability 107, 109n1
tips (informal experiments) 10–11
treatments. *see* independent variables
Tukey's HSD test and Scheffé's test 188
Two-factor ANOVA **189**, 193–94, 198–99
two-tailed hypothesis 138, 164, 176

two-thirds rule of thumb. *see* the 68% range
two-way chi-square tests 207–8, 210n3
type I errors 142–44, **145**
type II errors 143, 144, **145**

unbiased random sampling 41
unbiased samples 29, 33–34, 45

variability **18**, 79–80, 82, 83n1, 85–87, *see also* interquartile range (IQR); standard deviations
variable types 15–16, **18**; categorical data 16, **18**; continuous data 16, **18**; dichotomous variables 18, **18**; nominal 16, **18**; ordinal 16, **18**, *see also* scales of measurement
variables 15–17; interacting 195–97
variance 122, *123*, 127, 130; and the null hypothesis 157
variance accounted for 112, **124**, 125n2
vocabulary/reading exercise 122, *123*
volunteerism (self-selection bias) 28–29

Wilcoxon Matched-Pairs Signed Ranks Test 173n1
within-subject design (repeated-measures design) 170

x-bar symbol ($\bar{x}$) 70

z scores 91–92, 94; calculating 92–93
zero points 17, **18**